Structural Design of Interlocking Concrete Block Pavements

This book focuses on the structural design of Interlocking Concrete Block Pavement (ICBP) including application, analysis, design, construction, and the fundamental design concept of ICBP backed by case studies to demonstrate long-term performance of ICBP. It helps to draw conclusions regarding the various geometric properties that must be applied to achieve the performance of ICBP.

Features:

- Promotes the use of ICBP in the construction of pavements.
- Provides a comprehensive overview of concrete block paving.
- Discusses specifications from codes like ASCE 58-16, ASTM, BS and IRC codes and datasets online.
- Includes chapters on choosing the materials and designing the type and the shape of the concrete block and the thickness of all layers of the pavement.
- Lists appropriate stipulated values for the various geometric properties.
- Aids in understanding the history and comprehensive application of ICBP.

This book is aimed at graduate students and researchers in pavement and civil engineering.

Structural Design of Interlocking Concrete Block Pavements

Arjun Siva Rathan R.T and Sunitha V

CRC Press is an imprint of the
Taylor & Francis Group, an informa business

First edition published 2025
by CRC Press
2385 NW Executive Center Drive, Suite 320, Boca Raton FL 33431

and by CRC Press
4 Park Square, Milton Park, Abingdon, Oxon, OX14 4RN

CRC Press is an imprint of Taylor & Francis Group, LLC

ISBN: 9781032485799 (hbk)
ISBN: 9781032558134 (pbk)
ISBN: 9781003432371 (ebk)

DOI: 10.1201/9781003432371

Typeset in Times
by Newgen Publishing UK

Contents

Acronyms and Abbreviations

AASHTO	Association of State Highway and Transportation Officials
ASCE	American Society of Civil Engineers
BBD	Benkelman Beam Deflectometer
BIS	Bureau of Indian Standards
BPCF	British Precast Concrete Federation
BRO	Border Roads Organisation
CBGM+	Cement Bound Granular Mixtures
CBR	California Bearing Ratio
CMA	Concrete Manufacturers Association
CMAA	Concrete Masonry Association of Australia
CTB	Cement treated base
EAL	Equivalent Axle Load
EDA	Equivalent Design Axle
ESAL	Equivalent Single Axle Load
FAA	Federal Aviation Administration
FAARFIELD	Flexible Iterative Elastic Layer Design
FAAT	Fine Aggregate Angularity Test
FEA	Finite Element Analysis
FWD	Falling Weight Deflectometer
GSB	Granular Subbase
HKIA	Hong Kong International Airport
HVS	Heavy Vehicle Simulator
ICBP	Interlocking Concrete Block Pavement
ICPI	Interlocking Concrete Paver Institute
IPB	Interlocking Paver Blocks
IRC	Indian Road Congress
LVDT	Linear variable differential transformer
LWD	Light Weight Deflectometer
MDD	Maximum Dry Density
MEF	Material Equivalence Factor
MET	Method of Equivalent Thickness
MoRTH	Ministry of Road Transport and Highways
MSA	Million Standard Axle
OMC	Optimum Moisture Content
RTGC	Rubber Tied Gantry Cranes

SSR	Shear strength ratio
WBM	Water Bound Macadam
WMM	Wet Mix Macadam
YICT	Yantian International Container Terminals

Author Biographies

Arjun Siva Rathan R.T completed his M.S in Transportation Engineering in 2015 and Ph.D. in the field of Interlocking Concrete Block Pavement (ICPB) in 2022, from the National Institute of Technology Tiruchirappalli, India. He is currently working as Assistant Professor (Senior Grade) in the department of Civil Engineering, Amrita School of Engineering, Amrita Vishwa Vidyapeetham, Coimbatore. He has published 14 research articles in SCIE indexed journals and published nine international conferences. He is currently guiding six Ph.D. scholars. He is serving as a technical committee member for the reaffirmation of ASCE 58-21 guidelines on Structural Design of Interlocking Concrete Pavement for Municipal Streets and Roadways. He received best reviewer award from the *Journal of Cleaner Production* for the month of October 2018. He also received budding researcher award from the National Institute of Technology Tiruchirappalli. He is also a reviewer for several SCIE indexed journals. He is a member of professional bodies such as Indian Geotechnical Society, and Association of Consulting Civil Engineers. He has also undertaken several research projects and consultancies sponsored by private industries and Government agencies.

Sunitha V completed her B.Tech. in Civil Engineering from University of Calicut in 2001, M. Tech. in Traffic and Transportation Planning from NIT Calicut, India in 2003, and Ph.D. from IIT Madras, India in 2013. She is currently working as Associate Professor in National Institute of Technology, Tiruchirappalli. She has 20 years of experience in teaching and research. She has completed five sponsored projects worth Rs. 465 lakhs sponsored by Coir Board, Ministry of Urban Development, and IIT, Madras; NRIDA and MSME. She is the coordinator of State Technical Agency of Prime Minister Gram Sadak Yojana (PMGSY) program for Tamilnadu and Pondicherry. Her area of research interest is on pavement materials and pavement management systems. She has guided 83 postgraduate and 18 undergraduate project works in various fields such as Transportation Engineering, Transportation Planning, and Pavement Engineering. She also completed guiding four Ph.D. and two MSc. scholars and is currently guiding six Ph.D. scholars. She has published 90 papers in international/national journals and international/national conferences, out of which 28 journal publications are SCI/Scopus indexed and have one patent published. She is also involved in transportation related consultancy works. She is a member of professional bodies such as Indian Roads Congress, Institution of Engineers, Indian Geotechnical Society, ASCE, and TRG India.

Acknowledgements

It is a pleasant task to express our gratitude towards all those people who helped to complete the book. We would like to express our wholehearted thanks to the indomitable and invincible God for his blessings that are showered upon us in enabling us to complete our work. We also wish to express our deepest sense of gratitude to the professors, Dr. Moses Santhakumar, Dr. Samson Mathew, for their endless support in the form of expertise, assistance, and guidance throughout the process of this work. We are thankful to the Centre of Excellence in TRANSportation Engineering (CETransE), National Institute of Technology, Tiruchirappalli, for enabling us to complete this work by providing the necessary facilities. We are also thankful to Dr. Sasangan Ramanathan, Dean Engineering and Dr. Mini K.M, Chairperson, Department of Civil Engineering, Amrita School of Engineering, Coimbatore, Amrita Vishwa Vidyapeetham for their valuable guidance and support. Our sincere thanks are extended to all the faculty members in the Department of Civil Engineering, Amrita Vishwa Vidyapeetham, Coimbatore and National Institute of Technology Tiruchirappalli, whose inspiration has contributed immensely to accomplish this work. Our sincere thanks to our family members and friends individually for their continuous encouragement and motivation in hard times for the successful completion of the book.

Introduction

The Interlocking Concrete Block Pavement (ICBP) is a discrete pavement type used in light to heavy traffic areas. ICBP is said to be one of the most underrated pavement types, and the potential advantages of ICBP have not yet been fully acknowledged throughout the world. Only a few countries conduct considerable research into ICBP and promote ICBP research. The book on the structural design of ICBP intends to focus on the application, analysis, design, and construction of ICBP. The book chapters start from the history of ICBP to the design procedure of ICBP in roads, ports and airports followed by different countries. The book facilitates readers in comprehending the fundamental design concept of ICBP. The readers are also helped throughout the book to draw conclusions regarding the various geometric properties that must be applied in order to achieve the performance of ICBP. The appropriate stipulated values for the various geometric properties are listed in the book based on various literature and research findings. The reader and the researcher may find that the book is helpful in comprehending the structural analysis and interlocking mechanism of ICBP, in addition to the ways in which ICBP is distinct from conventional flexible pavements. The software used for the design of ICBP covered in this book is advantageous for the recent trends of research on ICBP. The case studies that are illustrated in this book will assist in demonstrating the long-term performance of ICBP. Finally, the book is intended to highlight the identity of ICBP in the road pavements, ports and airports and to promote the construction of ICBP roads.

Introduction to ICBP

1

1.1 INTRODUCTION

Roads are an essential component of every nation's infrastructure because of their contribution to a nation's overall economic growth and their ability to facilitate connectivity as well as the flow of people and products. Different types of pavements are used for road construction. Flexible and rigid pavements are the two types of pavements that are most frequently used for the construction of roads. Interlocking Concrete Block Pavement (ICBP) is yet another pavement type utilised extensively in various nations. The historical road design principle of stone interlocking served as the catalyst for the development of ICBP. Paver blocks are a common choice in pavement construction because of its fundamentally advantageous qualities, including high strength and durability, an aesthetically pleasing surface, easy maintenance, faster installation, and ease of removal, refit, and reuse. The ICBP is mostly employed in low-traffic regions such as residential roads, intersections, toll plazas, pedestrian crossing zones, parks, walkways, patios, parking bays, and high-volume roadways. Additionally, the ICBP has expanded its application in factories, container depots, gas stations, airports, and ports. However, the application of the ICBP is best suited for areas with traffic speed of 50 to 60 kmph. Even though the utilisation of ICBP is not comparable to that of traditional flexible and rigid pavement, the proportion of utilisation of ICBP has increased in comparison to previous years.

1.2 HISTORY OF ICBP

The ICBP is a historic, traditional pavement that emerged in the early 16th century. Ancient roman roads are constructed using stone pavers which replicate the hard surface to sustain military movements.[1] The wearing layer of the roman roads are laid using small-sized dressed stones such as cobbles, stone blocks of volcanic

DOI: 10.1201/9781003432371-1

tuff and stone from limestone and basalt. In 1622, a French lawyer named Bergier published the first work after his research from the remains of roman roads near Rheims, a city in France. The defined technology of stone pavement influenced and promoted the infrastructural development of France during the 17th and 18th centuries.[2] This infrastructural development in France is motivated by the British road expert Metcalf, Macadam and Telford in the 19th century. The present primary well established scientific approach in road construction is derived from these British experts based on stone pavements. The British road builders Metcalf, Macadam and Telford adopted a scientific approach in developing road construction techniques in the nineteenth century which are essentially unchanged today.[3] After the Second World War, Holland was the first country to introduce ICBP. Subsequently, this trend was followed by other countries across the world.[4] In 1929, the first British Standard was introduced with test methods for strength and water absorption, BS 368, being updated to BS 7263-1 and currently BS EN 1339. The potential application of the ICBP in the United Kingdom (U.K.) emerged in 1970s from the Cement and Concrete Association (C&CA).[5]

The flexible pavement used in practice nowadays are emerged during the second part of the 19th century. The better functional performance of the flexible pavement overtook the stone pavements all over the world. Meanwhile, the stone material in the wearing layer is converted into concrete material and named as Interlocking Concrete Block Pavement (ICBP). Due to the use of sustainable concrete material the research on the ICBP has emerged since 1980. The researchers namely Shackel B and Knapton J have made a significant contribution to the study of the ICBP. The research findings elevated the application of the ICBP from low volume roads to high traffic roadways, ports, and airport taxiways. In Chapter 7, case studies on the deployment of ICBP on heavily loaded locations that exhibit substantial performance are explained in detail. The foundation known as the Interlocking Concrete Paver Institute (ICPI) was established in the year 2000 with the primary purpose of conducting research and education on ICBP. The institute has made a significant contribution to the global growth of ICBP. They collaborated with the American Society of Civil Engineers (ASCE) to develop the ICBP guidelines ASCE 58 (2016), which are used worldwide. According to thc ICPI 2020 annual report, the global installation of ICBP increased by 56.57% between 2011 and 2019.[6] Today, ICBP is regarded as one of the popular and effective pavement types in the transportation sector.

1.3 APPLICATIONS OF ICBP

ICBP are utilised as a surfacing material for pavements all over the world. Moreover, ICBP are widely used and continue to gain popularity in Europe.[7] Interlocking Concrete Block Pavement (ICBP) is used in the pavement applications throughout

Australia and the world. In Australia, ICBP applications have been made for landscaping, canal linings, pedestrian areas, residential streets, urban roads and crossroads, bus terminals, airports, and industrial pavements since 1977.[8] ICBP is employed in low-volume areas such as residential roads, intersections, toll plazas, pedestrian crossing zones, parks, walkways, patios, parking bays. It is also used in heavy load areas such as container depots, industries, gasoline stations, airports, and ports. ICBP was laid for the apron and cargo loading area in the Hong Kong airport, covering 4,64,000 m^2 area.[4] ICBP are used in several industrial, governmental, and commercial applications today. Superior engineering properties, low maintenance, ease of placement and removal, reuse of original blocks, aesthetic appeal, and immediate use after installation or repair are primary reasons for selecting CBP over other paving surfaces.[9]

In hilly areas, the ICBP are currently employed successfully at hairpin bends. The use of ICBP is essential for safeguarding the roadways against landslides and snowfall. According to field research conducted in India, ICBP is a suitable solution to the problem of snowfall that exists on the Himalayan mountain ranges.[10] In India, the Border Roads Organisation (BRO) is constructing ICBP over snow-covered mountain routes at high altitudes. They discovered that the ICBP simplified snow removal operations utilising tracked earthmovers and decreased road damage. Also, this method has dramatically reduced the cost of maintenance and ensured that roads in difficult terrain are cleared quickly and with little disturbance to traffic. Also, the laying of the ICBP has improved the ride quality of the roads and decreased wear and tear on machinery and cars.[11] The ICBP is additionally utilised in embankments to stop soil erosion and strengthen them against slope failures. The ICBP constructions on the sloping sections of the road avoid horizontal creep and the ensuing opening of joints.[12] The application of ICBP is now gaining attention on all countries in different areas of pavement construction.

1.4 ADVANTAGES OF ICBP

The advantages of using ICBP for the road and airfield pavements are as follows:[13]

- ICBP is cost-effective as its low maintenance costs and high salvage value reduce the life cycle cost.
- ICBP is not adversely affected by fuel and oil spills, and it has higher resilience to the freeze and thaw effect.
- ICBP can withstand higher deflections without structural failure and is also resistant to thermal expansion and contraction.

- Precast blocks have a shorter production time and are more readily available than bituminous material.
- ICBP can be opened to traffic immediately after it is laid. Permitting traffic over the ICBP has the benefit of stiffening and interlocking the pavement.
- ICBP can be easily disassembled and reinstalled in the event of any subsurface services or for repairing the underlying supporting layers.
- Due to the fact that the blocks are manufactured in a factory, they are extremely high quality, thereby eliminating the challenges of quality control on the job site.
- The block pavements are an effective solution for intersections that require speed limits to be enforced and have high stress levels associated with turning.
- When trenches for utility repairs need to be dug, the process is simplified when the pavement is made of ICBP when compared to conventional flexible and rigid pavement.
- They are favoured in heavy-traffic places such as container terminals and ports because they are extremely well engineered to resist the high stresses.
- Because the concrete blocks are grey in colour, they are more visible at night than flexible pavement.
- Block pavement, in contrast to concrete pavements, does not show a significant detrimental effect from thermal expansion and contraction, and is not subject to the phenomenon of cracking.
- In cities and towns, the use of permeable block pavement can assist replace dwindling subsurface water supplies, filter contaminants before they reach open water sources, reduce storm water runoff, and lessen the need for drainage infrastructure.

Zero-slump Interlocking Paver Blocks (IPB) are appropriate for many commercial, municipal, and industrial applications due to their higher strength and durability. The high strength of IPB makes it suitable for use in container terminals; nevertheless, mostly they are being designed as flexible pavements, and their design may be over conservative in this regard.[13] From an aesthetic standpoint, the variety of block shapes and laying patterns offers a considerable deal of design flexibility. Moreover, the blocks are available in an array of colours. A variety of colours can be used as delineators to indicate the position of subsurface services, direction of traffic and the location of parking lots. Blocks can be used as delineators instead of paint, which also reduces maintenance expenses. IPB is an elegant building material that is perfect for use in housing developments, bus terminals, and pedestrian malls.[14,15] It is both economically and aesthetically advantageous to mix paving blocks and paving flags for some purposes, such as footways and pedestrian precincts, providing even more flexibility in the future to develop new and unique surface patterns in locations like pedestrian precincts.[16]

1.5 LIMITATIONS OF ICBP

Despite the numerous benefits of ICBP, there are a few drawbacks that are outlined below.[13]

- High speed facilities like expressways and national highways cannot employ concrete block pavements.
- The riding quality is adequate for low-speed traffic but inferior to that of a machine-laid bituminous or concrete pavement.
- ICBP may subjected to the growth of weeds if not frequently used.
- The ICBP produces more noise during the vehicle movement when compared to the conventional flexible and rigid pavement.
- It is necessary to pay careful attention to the drainage of the pavement since water can infiltrate through the joints.

The main constraint of ICBP is that it is best suited for locations where the design speed is less than 50 kmph. The main reason is that the end profile of the ICBP wearing layer is not smooth and continuous like that of flexible and rigid pavement. The presence of joints in ICBP causes an uncomfortable driving experience when the vehicle speed exceeds 50 kmph. In addition, the joints are responsible for a significant percentage of the noise that is produced by the movement of the vehicle. No experiments have been performed to precisely ascertain the rate at which the noise level increases to unacceptable levels. Another significant issue faced by ICBP is decreased skid resistance, which is mostly attributable to the manufacturing process of IPB. The IPB produced using a dry process has a rough surface that offers improved skid resistance. The surface of the IPB produced via a wet process is smooth and has a lower coefficient of friction. The drainage and stabilisation of jointing sand are two more significant drawbacks. The sealing and stabilisation of ICBP joints will consequently be effective for heavy industrial applications of block paving, such as ports, airports, and gas stations. In the case of airport pavements, stability of jointing sand is deemed necessary to prevent damage to aircraft engines caused by sand ingestion.[17] For ICBP to be more functional, it is necessary to address limitations such as the negative effects of poor drainage while providing the unbound base layer and the provision of the suitable gradation of bedding and jointing sand. The cracking of cement-stabilized layers and the probable loss of sand into these cracks constituted a potential issue. In this situation, using a primer seal to cover the base layers would be ineffective because the seal would also crack. Geotextiles plays more effective role in reducing the sand particle intrusion.[18] The constraints on the implementation of ICBP may be overcome with the aid of research in these areas.

1.6 STRUCTURAL DESIGN OF ICBP

The structural design of ICBP is complicated due to the presence of discrete wearing layer. Every country has its own design methodology and guidelines. Most countries established ICBP designs based on flexible pavement design guidelines, assuming that the ICBP functioned similarly to flexible pavement except for the wearing layer. Like those in flexible pavement, failures in the supporting layers will be reflected in the wearing layer of ICBP. However, the transmission of applied load to the supporting layer of ICBP differs from that of flexible pavement. This is due to the lateral load dissipation ability of the ICBP wearing layer as a result of the improved interlocking. In general, block pavements tend to perform in a manner which is qualitatively similar to the conventional flexible pavement. The Interlocking Concrete Paver Institute (ICPI) is a research institute in United States and Canada that specifically engages in technical, educational and Governmental activities of ICBP. The American Society of Civil Engineers (ASCE) in collaboration with the Interlocking Concrete Paver Institute, created the exclusive guidelines for ICBP as ASCE 58-16. In Australia, the Concrete Masonry Association of Australia (CMAA) established guidelines for the design and construction of ICBP. The guidelines provide information on the design of concrete segmental pavements for residential accessways, but the design guidelines for the ICBP for heavy-duty and domestic driveways were not considered as the scope of the guidelines. The Concrete Manufacturers Association (CMA) created the ICBP guidelines for South Africa, primarily focusing on ICBP preparation, compaction, and construction method. A number of design methods have been developed for ICBP which includes domestic applications, roads, heavy duty port container areas, and airport pavements.[19] The detailed design guidelines provided by different countries are discussed in Chapter 7.

The premise of ICBP was initially established in India in the early 1990s. Panda (2000) stated that the potential benefits of ICBP have not yet been wholly recognised in India since there are no design charts and construction information on this pavement. Later in 2004, the Indian Road Congress issued IRC SP 63 (2004) as the standard for ICBP, including the design catalogue and construction procedure. In India, the ICBP guidelines are provided by the Indian Road Congress (IRC) as IRC SP 63 (2018), while the IPB specification is developed by the Bureau of Indian Standards (BIS) as IS 15658 (2006). The flexible pavement was designed according to IRC 37 (2018) guidelines, which incorporate fatigue and rutting as failure criteria for flexible pavement design.

As a result, the flexible pavement is made to endure fatigue cracking caused by horizontal tensile strain and rutting caused by vertical compressive strain. IRC SP 63 (2018) specifies that the performance of ICBP is comparable to a flexible pavement, except the wearing layer. This could result from the discontinuous wearing layer reflecting the failure of the underlying layers. However, the grain-to-grain stress transfer of flexible pavement does not apply to ICBP. In general, block pavements tend to perform in a manner which is qualitatively similar to the conventional flexible pavement.

1.7 CONCLUSIONS

Roman roadways constructed with segmental stones served as the basis for the construction of ICBP. Due to the presence of a distinct wearing layer, the performance of ICBP is distinctive. The ICBP is a consistent pavement that encompasses all requirements imposed for traditional flexible and rigid pavement. Due to its discontinuous wearing layer, the application of ICBP is restricted to high speed expressways and highways. The distinguishing benefits, such as aesthetics, ease in replacement, minimum maintenance, etc., expanded the application of ICBP to a wide range of applications, from heavily trafficked port areas and airport pavements to rural roads. Due to the discrete wearing layer, the structural design of ICBP is intricate. The complexity of the ICBP design is further influenced by the use of various materials, such as Interlocking Paver Blocks (IPB), bedding sand and jointing sand. The other parameters of IPB such as shape, thickness, laying pattern and size also play a remarkable role in the structural performance of ICBP. The ICBP, aside from the wearing layer, is generally acknowledged to function as flexible pavement. Thus, the design of flexible pavement serves as a design guide for ICBP.

REFERENCES

1. Duducu, J. (2015) The Romans in 100 Facts. *Amberley Publishing*. ISBN 9781445649702.
2. Knapton, J. (1986) The Structural Design of Heavy Duty Concrete Block Pavements. *International Workshop on Interlocking Concrete Pavements*, Melbourne, Australia.
3. Karatag, H., S. Firat, and N. Işik (2017) Evaluation of Flexible Highway Embankment Under Repetitive Wheel Loading Using Finite Element Analysis. *Buildings Symposium (ISBS 2017)*, 1(3), 705–716.

4. ICPI (2004) Case Study in Engineered Interlocking Concrete Pavement Hong Kong International Airport – Report. *Interlocking Concrete Pavement Institute*, Virginia, United States.
5. Shackel, B. (1984) The Design of Interlocking Concrete Block Pavements for Road Traffic. *Second International Conference on Concrete Block Pavement,* Delft, The Netherlands.
6. ICPI (2020) Annual Report 2020. *Interlocking Concrete Pavement Institute*, Virginia, United States.
7. Molenaar, A.A. (1992) Concrete Block Pavements-European Practice. *Concrete Segmental Paving Workshop,* Perth, Australia.
8. Shackel, B. (1992) Australian Practice in Concrete Block Paving. *Concrete Segmental Paving Workshop*, Perth, Australia.
9. Ghafoori, N., and R. Mathis (1998) Prediction of freezing and thawing durability of concrete paving blocks. *Journal of Materials in Civil Engineering*, 10(1), 45–51.
10. Muraleedharan, T., and P.K. Nanda (1996) Laboratory and Field Study on Interlocking Concrete Block Pavement (ICBP) for Special Purpose Paving in India. *Fifth International Conference on Concrete Block Paving,* pp. 413–421, Tel Aviv, Israel.
11. https://timesofindia.indiatimes.com/city/pune/border-roads-organisation-blazes-a-trail-in-most-testing-conditions/articleshow/95608878.cms
12. Mudiyono, R., H.M. Nor, M.R. Hainin, and T.C. Ling (2006) Performance of concrete block pavement on sloped road section. *Doctoral dissertation*, Universiti Teknologi, Malaysia.
13. IRC SP 63 (2018) Guidelines for the use of Interlocking Concrete Block Pavement. *Indian Road Congress Special Publication*, New Delhi.
14. Sharp, K.G. (1979) Interlocking Concrete Block Pavements: Preliminary State-of-the-Art Review. *Australian Road Research Board Conference Proceedings*, Vermont South, Victoria, Australia.
15. Ghafoori, N., and R. Mathis (1997) Sulfate resistance of concrete pavers. *Journal of Materials in Civil Engineering*, 9(1), 35–40.
16. Lillcy, A. (1986) Concrete block paving in Great Britain by 1986. *International Workshop on Interlocking Concrete Pavements,* Melbourne, Australia.
17. Emery, J.A. (1993) Stabilization of jointing sand in block paving. *Journal of Transportation Engineering*, 119(1), 142–148.
18. Sharp, K.G., and P.J. Armstrong (1985) Interlocking Concrete Block Pavements, Special Report. *Australian Road Research Board,* ISBN 086910 196.
19. Vroombout, F., and N.G. Bentley (1992) Construction and Maintenance Aspects of Concrete Block Pavements for Heavy Aircraft Use. *Proceedings, 16th Australian Road Research Board Conference,* Perth, Western Australia, 16(2).

Pavement Materials

2

2.1 INTRODUCTION

The wearing layer of ICBP is discrete which comprises different materials including Interlocking Paver Block (IPB), bedding sand and jointing sand. The supporting layers consist of base layer, subbase layer and subgrade. The IPB is casted as precast blocks in the manufacturing industry. The main constituents in the manufacturing of IPB are cement, fine aggregate and coarse aggregate of nominal size 12.5 mm. There are two different casting processes in the manufacturing of namely dry process and wet process. The dry process involves casting the IPB using dry mix with low water to cement ratio and prepared using hydraulic pressing. The wet process involves high water-cement ratio with added admixture and hence requires low cost vibratory equipment. The base and subbase layer possess the specification of flexible pavement design as followed by the respective countries. The subgrade is the lowest supporting layer for the ICBP which should meet the minimum California Bearing Ratio (CBR) of 5%. All the materials should meet the desired properties to be adopted in the ICBP construction for better performance.

2.2 INTERLOCKING PAVER BLOCKS (IPB)

The IPB is the topmost wearing layer in the Interlocking Concrete Block Pavement. The IPB is made of concrete and manufactured in the casting industry.

DOI: 10.1201/9781003432371-2

2.2.1 Manufacturing of IPB

The manufacturing of IPB is carried out in two different processes namely dry and wet process. The manufacturing process is explained as follows.

(a) Dry Process

The dry process is a method of manufacturing the IPB through hydraulic press process as illustrated in Figure 2.1. The manufacturing of IPB using dry process is economical and requires less labour. The water-cement ratio for the concrete mix is kept low within a range of 0.3 to 0.4. The curing period in the dry process is found to be lower when compared to the wet process. The higher productivity is witnessed in the dry process when compared to the wet process. The IPB production using the dry process ranges from 2000 to 3000 blocks per hour. The processes involved in the dry method of manufacturing of IPB are illustrated in flowchart and pictorial representation.

Stage I: Raw Materials such as cement, aggregate and water stored in silos
Stage II: Batching of the required material through automatic batching controller
Stage III: Mixing of the raw materials based on batching
Stage IV: Transportation of the raw materials through a conveyor belt
Stage V: Hydraulic pressing after laying raw materials in a mould
Stage VI: Air curing at stackyard

(b) Wet Process

The wet process involves the placing of concrete in a rubber mould without hydraulic pressing. The IPB is manufactured with higher water content with the use of a superplasticer and viscosity modifier. The high workability concrete is laid on the rubber mould placed over a vibratory table. The vibratory table aids in possessing better compaction without segregation. After the compaction for 2 to 3 minutes, the rubber mould is removed after 8 to 10 hours and kept for air and subsequent water curing. The casting rate of the IPB depends on the size of the vibrating table and the number of moulds. The IPB manufactured through the wet process ranges from 500–1000 blocks per hour. The dry process is a quick process as it is automated and machine based. The long term studies highlight the fact that the IPB manufactured in the wet process is costlier when compared to the dry process. However, the initial procurement cost of the hydraulic pressing machine used in the dry process is higher. The process involved in the wet process are as follows:

FIGURE 2.1 Dry process – Methodology (Bharath Pavers, Madurai).

Stage I: Raw materials such as cement, aggregate and water are manually batched
Stage II: Mixing of the raw materials based on batching on rotating drum
Stage III: Placing of the rotating drum over the rubber mould in vibratory table
Stage IV: Compaction of the placed concrete on a vibratory machine
Stage V: Removal of IPB from rubber mould
Stage VI: Curing at stackyard

When compared to dry process, the wet process is time consuming and requires labour requirement. The dry process provides a higher strength when compared to the wet process. However, the initial cost of the hydraulic pressing machine is high which cannot be afforded by small scale industries. The wet process requires a vibratory table and rubber mould, which is low cost equipment and affordable for small scale industries which enhances the production cost. Another major concern is that the surface of the IPB manufactured through wet process is smooth which reduces the coefficient of friction. Therefore, for heavy traffic areas the IPB manufactured through the dry process is preferred compared to the wet process.

2.2.2 Quality Testing of IPB

The testing of IPB is required for the assurance on the quality of IPB. There are different codes followed by different countries. The testing of IPB is discussed below in Table 2.1.

2.2.3 Desired IPB Properties

The desired property for IPB properties is provided in Table 2.2, based on different codal specifications:

2.3 BEDDING SAND AND JOINTING SAND

Natural river sand is used as the bedding and jointing sand for the construction of ICBP. Nowadays, due to the scarcity in natural resources, the manufacturing sand is used as bedding and jointing sand. Panda and Ghosh[1] claimed that the M-sand is considered an effective source material for the bedding sand and

TABLE 2.1 Tests on IPB

SL. NO.	*DESIRED PROPERTY*	*METHODOLOGY AND FORMULA*	*CODE*
1.	Plan Area	The IPB's are manufactured in various size and shapes. It is required to determine the surface area for different shape. The plan area test helps in determining the plan area of any irregular shapes used as IPB. The IPB to be measured is laid over the cardboard of uniform size and trace around its perimeter with a pencil. The traced line is cut out accurately and measured in the weighing balance to the nearest 0.01g. A standard rectangular shape of size measuring 200mm × 100mm is cut into a same size. The plan area is calculated based on the equation 1 $A_s = \frac{20000\, m_s}{m_r}$ *Where,* m_s *is the mass of the cardboard shape matching the block (in g);* m_r *is the mass of the 200 mm × 100 mm cardboard rectangle (in g);* *or by using other means capable of measuring to 10 mm*^{2}*; and* *or by using the manufacturer's declared value.*	BIS 15658[2] & BS 6717 – 1[3]

(***continued***)

TABLE 2.1 (Continued)

SL. NO.	*DESIRED PROPERTY*	*METHODOLOGY AND FORMULA*	*CODE*
2.	Water Absorption (%)	The water absorption of IPB is carried out by immersing the IPB at room temperature for 24 ± 2 h. After 24 h, the IPB is removed from the water and drained for 1min, the visible water in the surface of block is removed. The specimen is then weighed and noted as Ww. Subsequent to saturation, the IPB is dried in a ventilated oven at 107 + 7°C for not less than 24 h. The dry IPB is then measured in N or Kg. The percentage of water absorption is calculated as follows: $w\ percent = \frac{W_w - W_d}{W_d}$ *Where,* W_w = *Weight of saturated specimen and* W_d = *Weight of dry specimen in Newton or Kilograms;*	IRC SP 63[4]
3.	Compressive Strength	The compressive strength is a important parameter which simulates the strength of IPB against the traffic without cracking. The test is carried out by placing IPB in the compression apparatus. The constant load at the rate of 15 ± 3 MPa/Min. The correction factor needs to be applied based on the type of block and the maximum load at failure is measured and noted. The stress is calculated as per Eqn. The correction factor is shown below:.	

THICKNESS OF BLOCK (MM)	CORRECTION FACTOR	
	PLAIN BLOCK	CHAMFERED BLOCK
60 or 65	1.00	1.06
80	1.12	1.18
100	1.18	1.24

$$\sigma = \frac{P}{A}$$

Where,

σ = Ultimate stress at failure in MPa or kg/cm^2;

P = Ultimate load at failure in Newton or Kilograms; and

A = Area perpendicular to the applied load, in mm^2 or cm^2.

BS 6717 – 1,[3] SNI 03 – 0691[5]

4. Split Tensile Strength

The failure pattern witnessed in the compressive strength in field is different from that of the laboratory testing. The IPB is placed on the compression testing machine. A rod is kept along the longest splitting section with side face with at least 0.5 times the thickness of the IPB on the paver block over at least 75 percent of splitting section area. The IPB fail along the length of the applied load. The split tensile strength is calculated as follows:

$$T = 0.637 * k * \frac{P}{S}$$

BIS 15658,[2] BS 6717,[3] BS EN 1338[6]

(continued)

TABLE 2.1 (Continued)

SL. NO.	DESIRED PROPERTY	METHODOLOGY AND FORMULA	CODE
		Where, *T = Paving block strength in MPa;* *P = Measured load at failure in Newton;* *S = The area of failure plane in* mm^2*, equal to l * t;* *l = The mean of two measured failure lengths, in mm;* *t = Paving block thickness at the failure plane; and* *k = Correcting factor (see table).*	
5.	Flexural Strength	The flexural strength test for IPB is determined using two pointing loading. The diameter of the supporting rollers ranges from 25 to 40mm. The distance from centre-to-centre is provided as the length of the specimen minus 50mm. The load is applied with a constant rate of 6 kN/min and the flexural strength is calculated as $F_b = \frac{3Pl}{2bd^2}$ Where, F_b *= Flexural strength, in* N/mm^2*;* *P = Maximum load, in N;* *l = Distance between central lines of supporting rollers, in mm;* *b = Average width of block, measured from both faces of the specimen, in mm; and* *d = Average thickness, measured from both ends of the fracture line, in mm.*	BIS 15658,[2] ASTM D790,[7] BIS 516[8]

6.	Abrasion	The square shaped specimen measuring 71.0 ± 0.5 mm is cored from the IPB and placed in the abrasion machine with the testing face at the top. The abrasion is simulated with the rotating grinding disc with 30 RPM. The specimen is reoriented to ninety degree after the completion of every 22 revolutions which is considered as one cycle. The volume loss on abrasion is calculated after 16 cycles from the following equation: $\Delta V = \frac{\Delta m}{PR}$ *Where,* *ΔV = Loss in volume after 16 cycle, in mm^3;* *Δm = Loss in mass after 16 cycles, in g; and* *PR = density of the specimen, or in the case of two-layer specimens, the density of the wearing layer, in g/mm^3.*	ASTM C779[9]
7.	Freeze and Thaw	The freeze and thaw study aids in understanding the behaviour of IPB for adverse climatic conditions. The duration of the freeze thaw cycle is completed every 24 hrs. The cycle comprises of 16 ± 1 h of freezing, followed by 8 ±1.h of thawing. The test procedure is followed until 50 freeze-thaw cycles. The test can be stopped when the test specimen disintegrated more than 1% of the dry weight of the specimen measured before the commencement of the test.	BIS 15658[2]

TABLE 2.2 Desired values of IPB

SL NO.	*DESIRED PROPERTY*	*SPECIFIC VALUES*
1.	Water Absorption (%)	The average of three units shall not be more than 6 percent by mass and in individual samples, the water absorption should be restricted to 7 percent.
2.	Compressive Strength	The minimum compressive strength for the ICBP is M35. The compressive strength is provided up to for traffic
3.	Split Tensile Strength	The minimum compressive strength for the ICBP is 3.5 MPa.
4.	Flexural Strength	Minimum-Breaking Load (kN): Residential pathways/Public Pedestrian – 2 Residential driveways – Light Vehicles – 3 Residential driveways – Commercial Vehicles – 5 Regularly trafficked roads – 6 Heavy duty/industrial roads – 7
5.	Abrasion	The depth of wear after 60 minutes should not be more than 1.5mm in case of M30 grade block, 1.25mm in case of M40 grade block and 1mm in case of M 50 grade block
6.	Freeze and Thaw	The average weight loss of three paver blocks, after having been subjected to 50 freeze-thaw cycles while totally immersed in a 3 percent sodium chloride solution, shall not exceed 1.00 percent of the initial constant dry weight of the specimens,

jointing sand due to the presence of angular particles. The bedding sand is found to be coarse gradation and the jointing sand is a finer gradation. The finer gradation is preferred for jointing sand due to the fact that the coarser gradation may clog between the blocks and may not be filled completely. This may reduce the lateral load dissipation behaviour of ICBP. A few of the tests to be conducted for the quality control of the bedding sand and jointing sand are illustrated in Table 2.3 and Table 2.4. The gradation of bedding sand and the jointing sand are discussed in Chapter 3.

2.4 BASE AND SUBBASE

The base and subbase layers act as a separation and drainage layer of ICBP. The different base and subbase layers are listed in Table 2.5 with the testing procedure and desired properties.

2.5 SUBGRADE

The main function of the sub grade is to give adequate support to the pavement and for this the sub grade should possess sufficient stability under adverse climatic and loading conditions. The different properties of the subgrade are enlisted in Table 2.9.

2.6 SUMMARY

The materials to be considered for the design of ICBP are IPB, bedding sand, jointing sand and supporting layers. The supporting layer includes base, subbase layer and subgrade. Different tests are required to be conducted to determine the quality of the material to be used for the construction of the IPB. The test aims to improve the performance of ICBP. The testing and specification of the supporting layers of ICBP are similar to that of the flexible pavement. The minimum desired values based on the specification of various code provisions should be ensured during the construction of ICBP.

TABLE 2.3 Tests for bedding and jointing sand

SL NO.	*DESIRED PROPERTY*	*METHODOLOGY*	*CODE*
1.	Angularity and Particle shape	The angularity of fine aggregate is indirectly measured using the uncompacted void content determined using Fine Aggregate Angularity Test (FAA). The angularity is proportional to the measured uncompacted void content. The test is carried out by providing fine aggregate of definite grading in a 100ml cylinder and allowed to flow through a funnel from a nominal height. The fine aggregate is then weighed to nearest N. the uncompacted air voids are calculated as follows: $U=\frac{V-\left(\frac{F}{G}\right)*100}{V}$ *Where:* • *U = Percent uncompacted air voids;* • *V = Volume of the cylindrical measure (ml);* • *F = Sample mass (grams); and* • *G = Dry bulk specific gravity of the fine aggregate.*	AASHTO T304,[10] ASTM C 1252[11]
2.	Gradation test	The bedding sand for testing is dried at a temperature of 110 ± 50C and weighed to the nearest 0.1 percent. The dried sand is sieved using the standard sieve set. The material retained on individual sieve needs to be removed and weighed. For the jointing sand, the test sample after being dried and weighed shall be placed in the container and sufficient water added and agitated until the coarse particles finer than 75-micron gets removed. The amount of material passing the 75-micron IS Sieve shall be calculated as follows:	ASTM C136,[12] BIS 2386,[13] ASTM C117[14]

		$A = \frac{B-C}{B} * 100$ *Where:* • *A = Percentage of material finer than 75-micron;* • *B = Original dry weight; and* • *C = Dry weight after washing.*	
3.	Specific gravity test	The specific gravity of the bedding sand and jointing sand can be tested using Pycnometer test. The sand of 500g is placed in a tray and sample is saturated and weighed (Weight A). The sample is filled with the soil and water in pycnometer and measured (Weight B). The sample is removed from the pycnometer and is filled with water (weight C). The saturated sample is kept in oven at 100 to 110°C for 24 ± ½ hours and weighed (Weight D). $Specific\,Gravity = \frac{D}{A-(B-C)}$ $Apparent\,Specific\,Gravity = \frac{D}{D-(B-C)}$ $Water\,absorption(percent\,of\,dry\,weight) = \frac{100(A-D)}{D}$ *Where:* • *A = Weight in g of saturated surface-dry sample;* • *B = Weight in g of pycnometer or gas jar containing sample and filled with distilled water;* • *C = Weight in g of pycnometer or gas jar filled with distilled water only; and* • *D = Weight in g of oven-dried sample.*	ASTM C128,[15] BIS 2386[13]

(***continued***)

TABLE 2.3 (Continued)

SL NO.	*DESIRED PROPERTY*	*METHODOLOGY*	*CODE*
4.	Abrasion test	The abrasion test of fine aggregate is carried out using Micro Dovel test apparatus. A magnetic stainless steel balls of 1250 ± 5g is kept in the Micro-Deval container. Add 500 g of sand after properly washed in water and dried in the oven. Add 750 ± 50 mL of tap water at a temperature of 68 ± 9°F to the Micro-Deval container with the steel balls. The sample is kept idle for a minimum of 1 hr. Set the timer on the Micro-Deval machine to 15 minutes, record this time as T, record the number of revolutions registered by the tachometer as N. $RPM = \frac{N}{T}$ *Where:* *N = Number of revolutions registered by the tachometer of the Micro-Deval machine; and* *T = Time set on the timer of the Micro-Deval Machine, minutes.* *Calculate the Micro-Deval abrasion loss:* $Percent\ Loss = \frac{(A-B)}{A} * 100$ *Where:* *A = Initial dry weight of the test sample before testing, g; and* *B = Final dry weight of the test sample after testing, g*	Tex-461-A,[16] CSA A23[17]

5.	Constant head permeability	The oven dried sand of 2.5 kg is taken for the test. The constant head test permeameter mould is weighed and the specimen is placed and connected through inlet into the constant water reservoir. The outlet of the bottom of the mould is kept open. The geometry of the specimen mould is noted. The quantity of water collected at a convenient interval of time is noted. The height H_1 and H_2 are measured to determine the head loss h. the temperature of water T is also measured and recorded. The permeability of sand is calculated as follows: $k_T = \frac{Q}{Ait}$ $k_{27} = k_T \frac{\gamma_T}{\gamma_{27}}$ *Where,* k_{27} = *permeability at 27°C;* γ_T = *Coefficient of Viscosity at T°C;* γ_{27} = *Coefficient of Viscosity at 27°C;* Q = *Quantity in cm³;* A = *Area of specimen in cm³;* i = *hydraulic gradient; and* t = *time in seconds.*	ASTM D2434,[18] BIS 2720-17[19]

(*continued*)

TABLE 2.3 (Continued)

SL NO.	*DESIRED PROPERTY*	*METHODOLOGY*	*CODE*
6.	Shear test	The direct shear test is a laboratory testing method used to determine the shear strength parameters of bedding and jointing sand. The commonly used shar box size is 60mm × 60mm. Place moist porous inserts over the exposed ends of the specimen in the shear box. Cohesionless soils may be tamped in the shear box itself with the base plate and grid plate or porous stone as required in place at the bottom of the box. The normal stress is applied after providing porous stone on the specimen. The shear load is applied and the respective displacements are measured. The corrected area is determined as follows: *Corrected Area* $= A_0 (1-(\delta/3))$ *Where:* A_0 = *Initial area of the specimen in* cm^2 *and* δ = *displacement in cm.* The shear stress versus horizontal displacement is plotted. The maximum value of shear stress is read if failure has occurred, otherwise read the shear stress at 20% shear strain. The maximum shear stress versus the corresponding normal stress is plotted for each test, the cohesion and the angle of shearing resistance of the soil is determined from the graph.	ASTM D3080,[20] IS 2720-13[21]

TABLE 2.4 Desired properties for bedding and jointing sand

SL NO.	*DESIRED PROPERTY*	*SPECIFIC VALUES*
1.	Angularity and Particle shape	A minimum uncompacted void content of 45 percent is generally recommended for the blend of fine aggregate for a high traffic pavement. Minimum 60% combined sub-angular and sub-rounded.
2.	Gradation test	Maximum 1–5% passing through 75 microns for bedding sand and 5–10% for jointing sand.
3.	Specific gravity test	The specific gravity of the bedding sand is recommended to be between 2.6 to 2.8 and the specific gravity of the jointing sand is 2.3 to 2.6.
4.	Abrasion test	The abrasion loss of the bedding sand and joining sand should not exceed 8%.
5.	Constant head permeability	Minimum permeability of the bedding and jointing sand is 2 × 10-3 cm/second.
6.	Shear test	A friction angle of greater than 40° would be desirable to minimize rutting.[22]

TABLE 2.5 Types and desired properties for base and subbase layers

BASE/SUB-BASE LAYER	*MATERIALS*	*PROPERTIES*	*DESIRED VALUE*	*REFERENCE CODE*
Granular Layer	The material to be used for the work shall be natural sand, crushed gravel, crushed stone, crushed slag, or combination thereof depending upon the grading required. Use of materials like brick metal, Kankar and crushed concrete shall be permitted in the lower sub-base.	Gradation	Gradings III and IV shall preferably be used in lower sub-base. Gradings V and VI shall be used as a sub-base-cum-drainage layer as shown in Table 2.6.	MoRTH Specification[23]/ IS 2386[13]
		Water Absorption	Should not be more than 2 percent	IS 2386[13]
		Aggregate impact value	40 Max	IS 2386[13]
		Liquid limit (%)	25 Max	MoRTH Specification[23] & IS 2720[19]
		Plasticity Index (%)	6 Max	
		Field Density	98% dry density	
		CBR at 98% dry density	Min 30	
Lime/Flyash/ Cement Stabilized layer	Dry lime slaked at site or pre-slaked lime.	Purity -lime	Should not be less than 70% by weight of Quick lime (CaO)	BIS 1514[24]
		Fineness	Specific surface area should be minimum 250 sqm/kg	IRC SP 89[25] /MoRTH specification[23]
		Gradation	Maximum particle retained on 45 micron sieve is 40%	
		Lime reactivity	Minimum reactivity is 3.5 N/mm^2	

		Soundness	Autoclave – 0.8mm (Max), Le-Chatelier – 10mm (Max)	
		7 days unconfined Compressive Strength (MPa	Ranges from 1.5 MPa to 12 MPa	IS 2720,[19] IRC SP 89[25]/ MoRTH specification[23]
Water Bound Macadam (WBM)	Clean, crushed coarse aggregates interlocked by rolling, and voids thereof filled with screening and binding material with the assistance of water,	Gradation Minimum thickness (mm) Los Angeles Abrasion Value Aggregate Impact value Combined flakiness and elongation Index	As shown in Table 2.7. 75 Maximum of 40% Maximum 30% Maximum 35%	IRC 19[26] / MoRTH Specifications[23]
Wet Mix Macadam (WMM)	Clean, crushed graded aggregates and granular material mixed with water and rolled to a dense mass on the prepared surface.	Water Absorption Gradation – Table Minimum thickness (mm) Los Angeles Abrasion Value Aggregate Impact value Combined flakiness and elongation Index	No greater than 2% As shown in Table 2.8 Not less than 75 Maximum of 40% Maximum 30% Maximum 35%	IRC 109[27]/ MoRTH Specifications[23]

TABLE 2.6 Grading for granular sub-base materials

	PERCENT BY WEIGHT PASSING THE IS SIEVE					
IS SIEVE DESIGNATION	*GRADING I*	*GRADING II*	*GRADING III*	*GRADING IV*	*GRADING V*	*GRADING VI*
75.0 mm	100	-	-	-	100	-
53.0 mm	80–100	100	100	100	80–100	100
26.5 mm	55–90	70–100	55–75	50–80	55–90	75–100
9.50 mm	35–65	50–80	-	-	35–65	55–75
4.75 mm	25–55	40–65	10–30	15–35	25–50	30–65
2.36 mm	20–40	30–50	-	-	10–20	10–25
0.85 mm	-	-	-	-	2–10	-
0.425 mm	10–15	10–15	-	-	0–5	0–8
0.075 mm	<5	<5	<5	<5	-	0–3

TABLE 2.7 Grading for Wet Bound Macadam (WBM)

GRADING NO.	*SIZE RANGE*	*IS SIEVE DESIGNATION*	*PERCENT BY WEIGHT PASSING*
I	63 mm to 45 mm	75 mm	100
		63 mm	90–100
		53 mm	25–75
		45 mm	0–15
		22.4 mm	0–5
II	53 mm to 22.4 mm	63 mm	100
		53 mm	95–100
		45 mm	65–90
		22.4 mm	0–10
		11.2 mm	0–5

TABLE 2.8 Grading for Wet Mix Macadam (WMM)

IS SIEVE DESIGNATION	*53.00 MM*	*45.00 MM*	*22.40 MM*	*11.20 MM*	*4.75 MM*	*2.36 MM*	*600 MICRON*	*75 MICRONS*
Percent by Weight Passing	100	95–100	60–80	40–60	25–40	15–30	6–18	4–8

TABLE 2.9 Properties of subgrade

LAYER	*MATERIALS*	*PROPERTIES*	*DESIRED VALUE*	*REFERENCE CODE*
Subgrade	Materials used in subgrades are existing soil, moorum, reclaimed material from pavement, fly ash, pond ash etc.	Liquid Limit	Should not exceed 50	MoRTH Specification[23] / IRC36[28]
		Plasticity Index	Should not exceed 25	
		Free Swelling Index	Should not exceed 50%	
		Thickness to be compacted	Should not be less than 500mm	
		California Bearing Ratio (CBR)	Should not be less than 5%	
		Dry Unit Weight	Should not be less than 15.2 kN/m^3	

REFERENCES

1. Panda, B.C., and A.K. Ghosh (2001) Source of jointing sand for concrete block pavement. *Journal of Materials in Civil Engineering,* 13(3), 235–237.
2. BIS 15658 (2006) Precast Concrete Blocks for Paving – Specification. *Bureau of Indian Standards,* New Delhi.
3. BS 6717 (1989) Precast Concrete Paving Blocks. *British Standard Institution*, London.
4. IRC SP 63 (2018) Guidelines for the use of Interlocking Concrete Block Pavement. *Indian Road Congress Special Publication*, New Delhi.
5. SNI 03-0691 (1996) Paving Block. *Indonesian National Standard*, Indonesia.
6. BS EN 1338 (2003) Concrete Paving Blocks. Requirements and Test Methods. *European Committee for Standardization*, Brussels.
7. ASTM D790 (2017) Standard Test Methods for Flexural Properties of Unreinforced and Reinforced Plastics and Electrical Insulating Materials. *American Society for Testing and Materials*, Pennsylvania, United States.
8. BIS 516 (1991) Methods of Tests for Strength of Concrete. *Bureau of Indian Standards,* New Delhi.
9. ASTM C779 (2012) Standard Test Method for Abrasion Resistance of Horizontal Concrete Surfaces. *American Society for Testing and Materials*, Pennsylvania, United States.
10. AASHTO T 304 (2012) Uncompacted Void Content of Fine Aggregate. *American Association of State Highway and Transportation Officials,* Washington.
11. ASTM C 1252 (2017) Standard Test Methods for Uncompacted Void Content of Fine Aggregate. *American Society for Testing and Materials*, Pennsylvania, United States.
12. ASTM C 136 (2019) Standard Test Method for Sieve Analysis of Fine and Coarse Aggregates. *American Society for Testing and Materials*, Pennsylvania, United States.
13. BIS 2386 (1963) Method of Test for Aggregates for Concrete for Determination of Specific Gravity, Void, Absorption and Bulking. Part III. *Bureau of Indian Standards.* New Delhi.
14. ASTM C117 (2017) Standard Test Method for Materials Finer than 75-um (No. 200) Sieve in Mineral Aggregates by Washing. American Society for Testing Materials, *American Society for Testing and Materials*, Pennsylvania, United States.

15. ASTM C128 (2015) Standard Test Method for Density, Relative Density (Specific Gravity), and Absorption of Fine Aggregate. *American Society for Testing and Materials*, Pennsylvania, United States.
16. Tex-461-A (2016) Micro-Deval Abrasion of Aggregate. *Texas Department of Transportation*, Austin.
17. CSA A23 (2019) Concrete Materials and Methods of Concrete Construction/Test Methods and Standard Practices for Concrete. *Canadian Standards Association*, Toronto.
18. ASTM D2434 (2022) Permeability of Granular Soils (Constant Head). *American Society for Testing and Materials*, Pennsylvania, United States.
19. BIS 2720 (1986) Methods of Test for Soils. *Bureau of Indian Standards*. New Delhi.
20. ASTM D3080 (2011) Standard Test Methods for Direct Shear Test of Soils under Consolidated Drained Conditions. *American Society for Testing and Materials*, Pennsylvania, United States.
21. BIS 2720-13 (1986) Methods of Test for Soils – Direct Shear Test. *Bureau of Indian Standards*, New Delhi.
22. Lilley A.A. (1980) A Review of Concrete Block Paving in the UK Over the Last Five Years. *First International Conference on Concrete Block Paving*, Newcastle, England.
23. MoRTH (2013) Specifications for Roads and Bridge Works. *Ministry of Road Transport and Highways*, New Delhi.
24. BIS 1514 (1990) Methods of Sampling and Test for Quicklime and Hydrated Lime. *Bureau of Indian Standards*, New Delhi.
25. IRC SP 89 (2018) Guidelines for Soil and Granular Material Stabilization Using Cement, Lime and Fly Ash. *Indian Road Congress Special Publication*, New Delhi.
26. IRC 19 (2005) Standard Specifications and Code of practice for Water Bound Macadam (WBM). *Indian Road Congress Special Publication*, New Delhi.
27. IRC 109 (2015) Guidelines for Wet Mix Macadam (WMM). *Indian Road Congress Special Publication*, New Delhi.
28. IRC 36 (2010) Recommended Practice for Construction of Earth. Embankments and Subgrade for Road Works. *Indian Road Congress Special Publication*, New Delhi.

Factors Affecting the Performance of ICBP

3

3.1 INTRODUCTION

The load-spreading capacity of a block layer is primarily determined by the interlocking and lock-up phenomenon. The term 'interlocking' describes the geometric connection between neighbour blocks. The lock-up mechanism develops in block pavements after construction as a result of traffic movement and weathering. After the lock-up criteria is achieved, the blocks behave more as a composite layer than as distinct units. The interlocking mechanism of the ICBP depends on the geometric properties of IPB. The following factors are considered while evaluating the performance of the ICBP: block thickness, block shape, block laying pattern, block laying angle, block strength, joint width, jointing sand gradation, bedding sand gradation, bedding sand thickness, granular layer thickness, and subgrade strength. Dentated-shaped blocks have a larger frictional side area than rectangular blocks, which helps to distribute the load more evenly and reduce stress. In comparison to the stretcher and basketweave laying patterns, the herringbone pattern is more resistant to horizontal movement. The laying angle of IPB has minimal effects on the structural behaviour of ICBP. The variation in the compressive strength of the IPB has less of an effect on the load transfer mechanism or deflection behaviour of the ICBP. The sand particles used for jointing sand should be of sufficient size to fill gaps without voids and be sturdy enough to transfer load. The ICBP performance is more adaptable to the coarser bedding sand gradation. The performance of the ICBP is significantly influenced by the wearing layer and the supporting layers of ICBP.

DOI: 10.1201/9781003432371-3

3.2 GEOMETRIC PARAMETERS OF IPB

3.2.1 Block Shape and Size

The shape and size of the IPBs are crucial attributes that govern the deflection behaviour of the ICBP. There are numerous registered patents pertaining to IPB block shape.[1,2] Selection of block shape for the ICBP pavement is complicated by the availability of these various block shapes. There are three distinct categories for IPBs: A, B, and C. According to Figure 3.1,[3] Category A denotes interlocking on all sides, Category B indicates interlocking on two sides, and Category C implies interlocking on no sides. Each category has blocks in a variety of shapes. The Category C block, for instance, consists of hexagonal, square, and rectangular blocks. The shape of the block should be designed to improve stress distribution properties and ease of placement.[4]

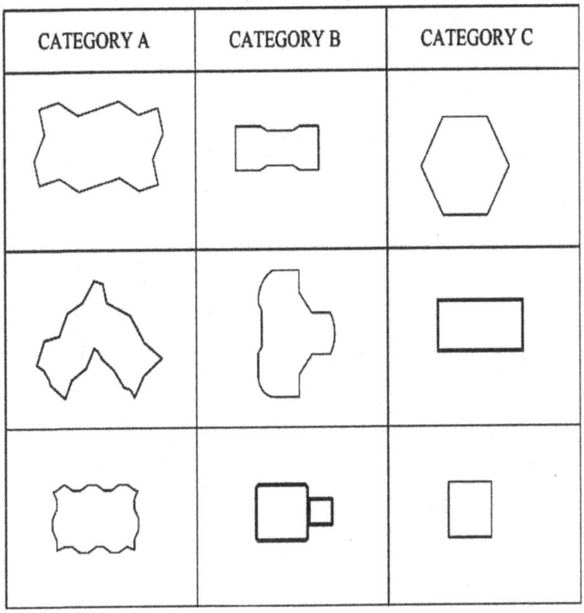

CATEGORY A	CATEGORY B	CATEGORY C

FIGURE 3.1 Different block shapes and sizes.

The block shape is one of the essential geometric properties of IPB that affects the interlocking mechanism of ICBP. It is also important to understand the deflection behaviour of the block shape with respect to the applied load. The block shapes are independent of load-distributing abilities and have no effect on the elastic deformation in the axis of load.[5,6,7] In contrast, literature studies claimed that block shape plays a major role in the structural behaviour of ICBP and also added that the modulus of ICBP section is affected by the block shape.[8,9,10] It was also postulated that the block shape has a substantial impact on the deflection of ICBP, modulus of ICBP and joint width.[8,11] These conclusions were based on the plate load testing conducted with blocks from three different categories. The Dutch universal design method for ICBP includes a block size of 105 mm × 211 mm as a critical parameter in the rutting prediction of ICBP.[12] It is found that the block shape plays a significant effect on the deflection behaviour of ICBP.

Most of the literature studies are carried out by considering blocks that fall under Category A and Category C i.e., zigzag-shaped and rectangular-shaped blocks. The zigzag block shape was considered as Category A, and rectangular blocks were considered as Category C. When subjected to extremely high vertical or horizontal loads, simple shaped Category C blocks are less prone to failure than complex profiled blocks that belong to category A type blocks.[13] The rectangular blocks are most suited for heavy industrial pavements due to advantages such as low stress concentration, ease of removal and uplift. In contrast, dentated and large sized Category A blocks have a higher frictional side area, which distributes more load and reduces stress when compared to rectangular blocks.[14] The field plate load test also determined that rectangular blocks tend to rotate, and I-shaped blocks are likely to shatter at the web-flange connection locations.[15] Compared to rectangular blocks, zigzag blocks or uni-shaped blocks provide better horizontal displacement resistance and reduced deflection when subjected to horizontal applied loading.[16,17] Zigzag shaped blocks and I shaped blocks show lower deflection and higher lateral resistance when compared to regular rectangular blocks. It is recommended that the block shape angle for the zigzag block is in the angle of 100°and 110° instead of 137° for better performance.[18] Literature study revealed that the majority of the study concluded the block shape as a critical parameter that influences the deflection behaviour of ICBP.

3.2.2 Block Thickness

Block thickness is another critical geometric component of IPB. Block thickness affects both the structural behaviour of the ICBP and the thickness of the supporting layers. Increase in block thickness results in an increase in the vertical surface area of the block. This increases the ability of the block to dissipate applied loads laterally, reducing the pressure on the base layers. Decreased pressure on the base layer results in lower deflection, which in turn reduces the thickness of the base layer. Thus, block thickness plays a major role on deciding the thickness of supporting layers such as base layer and subgrade. Increase in the block thickness reduces surface deflection, rutting deformation, subgrade stress and the stiffness of the supporting layers. The increase in the thickness of concrete blocks by 40mm lowers the thickness of Cement Treated Base (CTB) by 20mm. It was also found that when the block thickness increased from 60mm to 100mm, the rut depth decreased by 25% and the stress at the base layer was reduced by 30 to 65%.[17,19,20]

Few studies claimed that the block thickness has no substantial effect on the deformation of ICBP.[21,22] In contrast, it was noted that the increase in block thickness leads to lesser vertical stress, distorted pavement response and lower deflection. The decrease in deflection with regard to the increase in block thickness is nonlinear.[23] The nonlinear trend of the deflection due to the increase in the block thickness is due to the interlocking mechanism, which depends on the intensity of loading. When the loading intensity is higher, the interlocking occurs rapidly, resulting in higher lateral load dissipation.[24,25] The block thickness should be determined in accordance with the anticipated traffic loads, and the diameter of the loaded region is a pivotal factor that is directly proportionate to the block thickness. The jointing sand and joint width mainly influence the load dissipation of the block thickness as they act as a medium of load transfer. Based on the numerical simulations, it was found that when the joints are completely filled, the block thickness has a negligible effect on the deflection behaviour.[11]

Generally, a block thickness of 60mm to 120mm is preferred. The increase in the block thickness above 120mm is usually not recommended due to the difficulty in laying. A minimum block thickness of 60mm and maximum block thickness of 80mm is adequate for large volume traffic loads to sustain a design Equivalent Axle Load (EAL) of 2,000,000 repetitions.[26,27,28] The literature study revealed that block thickness plays a significant role in reducing the supporting layer thickness. It also plays a vital role in reducing the applied load to the base and subgrade layer.

3.2.3 Block Laying Pattern

Block laying pattern refers to the sequence and direction of laying blocks. The most commonly adopted laying patterns are herringbone, stretcher, and basketweave, as shown in Figure 3.2. Interestingly, the laying pattern is mostly determined by the block shape, as not all laying patterns are suitable for all block

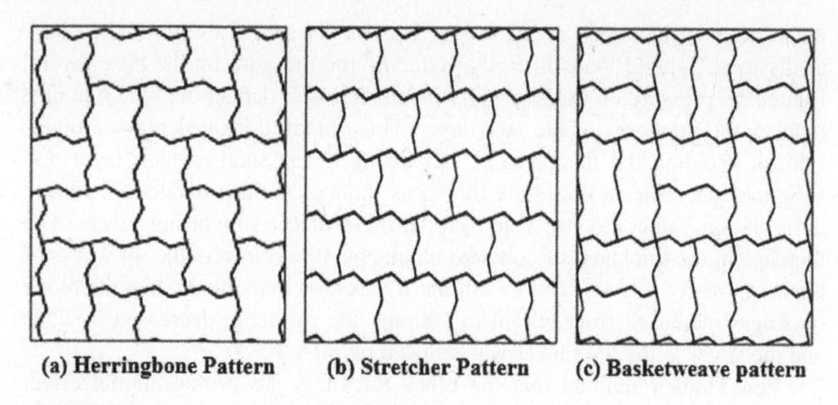

FIGURE 3.2 Different block laying pattern.

shapes. For example, an I shaped block cannot be laid using the herringbone pattern, implying that the block shape and block laying pattern are interrelated. Being a discrete layer, the IPB tends to have a horizontal movement during the application of traffic load. The laying pattern greatly influences the resistance to the horizontal movement. Few literatures claimed that the block laying pattern does not affect the structural behaviour of ICBP.[14,29,30] In contrast, it was reported that the laying pattern possesses a remarkable influence on the deflection behaviour of ICBP. The herringbone pattern has a greater D-moment and is more resistant to rotational and horizontal displacement.[16] The influence of laying pattern becomes adverse when the gaps between blocks continue to be parallel to the traffic direction.[31]

The stretcher and basketweave pattern have continuous parallel lines, whereas the herringbone pattern has a discontinuous line, as shown from Figure 3.2. These discontinuous laying lines of the herringbone pattern increase the resistance to horizontal movement and become more stable when subjected to horizontal load compared to stretcher and basketweave bonds. The herringbone pattern has lesser deflection and bending stress than the stretcher bond and is most suitable for high-load regions. The 90-degree laying procedure of the herringbone pattern is highly beneficial compared to other laying patterns, and this aids in lowering the horizontal displacement.[32,33] The herringbone pattern performs better under applied braking action due to improved interlocking efficiency, which reduces the vertical stress distribution of ICBP. In contrast to block shape and block thickness, the influence of the laying pattern on the deflection behaviour is widely accepted by the majority of researchers. However, all block shapes cannot be laid in a herringbone bond. The rectangular and zigzag shapes are the prominent block shapes best suited for a herringbone bond.

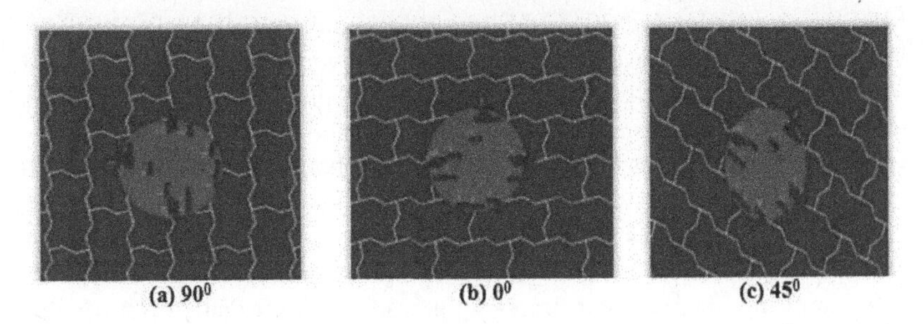

FIGURE 3.3 Block laying angle.

3.2.4 Block Laying Angle

The block laying angle is defined as the placing of IPB in different angles with respect to the traffic direction. The most common laying angles are zero, 45, and 90 degrees. The research on the effect of laying angle on the structural behaviour of ICBP is limited since parallel or perpendicular laying angles are generally preferred. According to the Concrete Masonry Association of Australia,[34] the performance of ICBP pavements when installed in a herringbone pattern at 90° or 45° does not significantly impact the structural behaviour of ICBP. When IPB is laid at 45^0, the patch loading and stress development are considerably increased, which is found to be in a reverse condition when the IPB is laid at 90^0. The block laying angles 0° and 90° are regarded as the optimal laying angles as they offer good load transfer.[35] Figure 3.3 represents the block laying angle used for the deflection study. An extensive literature review revealed that only a limited number of studies have investigated laying angle. The rationale for this could be that the herringbone bond is usually considered the best laying pattern due to its ability to minimise deflection compared to other laying patterns. The herringbone pattern is a combination of right-angle blocks that are laid both parallel and perpendicular to each other.

3.2.5 Block Strength

Compressive strength is a critical geometric property of IPB. IPBs are made of concrete, which is a mixture of cement, water, fine and coarse aggregates. Dry process is adopted for the manufacturing of ICBP. The rigid pavement should have a minimum flexural strength of 4.5 MPa. Thus, M40 is regarded the minimum grade of concrete for rigid pavement. However, in the case of the ICBP, the flexural strength is insignificant. The minimum grade of concrete is M30 grade for cycle tracks, and the maximum grade of concrete is

M50 for an axle load of 20–50 Million Standard Axle (MSA).[36,37] ASTM C 936 specifies IPB to have a minimum compressive strength of 55 MPa. Few literatures suggested that IPB should have a compressive strength greater than 60 MPa.[19] The compressive strength adopted for the IPB in the construction of ICBP varies in different countries. After examining the standards of various countries, it was concluded that the compressive strength used in various countries ranges from 25 MPa to 50 MPa.[38,39] The study on the usage of waste materials as a replacement of the conventional materials in the manufacturing of concrete is emerging nowadays. Ecofriendly high strength blocks are also made from byproducts such as granite waste and iron ore tailings.[40]

It is found that higher compressive strength provides better abrasion resistance compared to flexible and rigid pavements.[41,42] Furthermore, during heavy traffic, blocks with high strength transfer the permanent deformation to the supporting lower layers.[43] However, few studies concluded from the plate load test results of blocks with a compressive strength between 25 to 55MPa that the compressive strength had no effect on the deflection behaviour of ICBP.[1,9] This is because, despite having a higher compressive strength, the compressive stress on the block is lower due to its smaller size and discrete character. Also, the friction area, which is a major determinant in the load distribution of ICBP remains constant despite the variation in the compressive strength. Therefore, the load transfer mechanism or the deflection behaviour of ICBP has a lesser influence by the variation in the compressive strength of the IPB.

3.2.6 Edge Restraint

Edge restraints are installed along the edges of ICBP pavements to prevent lateral migration of blocks owing to traffic load, which disrupts the ICBP interlocking mechanism. The edge restraint are made with in situ concrete material above the subgrade and base layer.[44] The edge restraint is also made as precast material casted in industries and transported to the field. The general depth of the edge restraint is kept as 150mm and the width as 100mm.[45,46] A variety of materials, including concrete, timber, etc., can be used to provide edge restraint, but they must be rigid enough to sustain the lateral force that the pavement imparts on them. Prior to the construction of the pavement, the edge restraint should be installed. The blocks should be shielded from concrete staining by a sheet of polythene if an edge restraint concrete is poured after pavement construction since grout stains on the blocks are challenging to remove.[47] To provide a rigid footing for the edge restraint and to minimise swelling of expansive subgrade soils that could otherwise contribute to heaving or tilting of the edge restraint, the subbase construction should extend beneath the edge restraint over its entire width.[48] The joint gaps existing between the pavers and the edge restraint are filled using jointing sand with the percentage of silt and clay not more than 3%. The sand with 10% retained on the 5mm sieve is also used as the filling sand in joints.[45] The edge restraint plays a remarkable role in the reduction of deflection through enhanced interlocking between blocks. The interlocking mechanism developed due to the presence of edge restraint.[49] During application of vehicle traffic, the block tends to rotate which stimulates the horizontal movement of blocks. This lateral movement of blocks is restricted by the presence of edge restraint at the edge of the ICBP. This restriction in the horizontal and rotation movement forms a wedge and hinge formation which results in enhanced interlocking. The deflection of the ICBP is 1.6 times higher for pavements without edge restraints.[50] Thus, edge restraint plays a significant role in the performance of ICBP.

3.3 BEDDING SAND AND JOINTING SAND

3.3.1 Gradation of Bedding Sand

The gradation of the bedding sand plays a vital role in the performance and shear resistance of bedding sand. The stress pattern on the bedding sand depends on traffic conditions.[34] The stress created on the bedding sand was impacted mainly by IPB geometric parameters such as block shape, pattern, and orientation. The applied wheel load has a significant impact on the differential stress in the bedding layer as a result of eccentric load due to smaller size blocks. The thickness of the bedding sand was also found to be a critical parameter in the deflection of ICBP. After completing the field analyses at various airports, Knapton and Cook[51] concluded that bedding sand with finer particles resists water flow, increases hydrostatic pressure and decreases the shear strength of bedding sand. The gradation of bedding sand which follows Zone I and Zone II as per the specification of IS 383 possesses higher shear strength and lower deflection.[52] ICBP performs adversely when the bedding sand contains more than 15% of 75-micron sand particles.[1] Similarly, sand with a clay content of 20 to 30% exhibits 30mm deformation. The gradations recommended by different literatures are shown in Table 3.1.

TABLE 3.1 Different bedding sand gradations

SIEVE SIZE (MM)	*PERCENTAGE PASSING (%)*					
	ASTM C33[53]	*IRC SP 63*[2]	*SHACKEL*[54]	*GARRY*[55]	*PANDA AND GHOSH* [14]	*ARJUN ET AL.* [56]
9.5	100	100	100	100	100	100
4.75	95–100	95–100	95–100	60–100	94	100
2.36	80–100	80–100	80–100	60–90	70	88
1.18	50–85	50–95	50–85	25–70	46	72
0.600	25–60	25–60	25–60	10–35	30	45
0.300	5–30	10–30	10–30	5–20	16	21
0.150	0–10	0–15	5–15	0–10	6	6
0.075	0–1	0–10	-	0–5	2	2

TABLE 3.2 Different bedding sand thicknesses

SL NO	*DESCRIPTION*	*PROPOSED BEDDING SAND THICKNESS*
1.	ASCE 58 – 16 [59]	25mm before compaction
2.	Mavin[60]	30mm after Compaction
3.	Mujaj and Smith[46]	20–25mm
4.	IRC SP 63[61]	30 mm
5.	Simmons[62]	40mm
6.	Panda and Ghosh[14]	50mm
7.	Miura and Katsuhiko[58]	50mm in loose condition
8.	Arjun et al.[63]	30mm

3.3.2 Thickness of Bedding Sand

Bedding sand is a crucial component of ICBP. The primary role of the bedding sand is to provide adequate cushion to the IPB so that the wearing surface is uniform. Furthermore, the bedding sand acts as a barrier to propagate cracks from the base layers. The deflection behaviour of ICBP did not vary significantly with bedding sand thickness and shows negligible effect when the joints between blocks are completely filled.[11] In contrast, literatures reported that increasing the bedding sand increases the load-carrying capacity and interlocking efficiency.[57] A thin bedding layer reduces lock-up conditions because the vibration during the construction stage allows upward sand migration through the joints.[58] The different bedding sand thicknesses proposed in various studies are provided in Table 3.2. It was observed that the coarser gradation is more adaptive for the ICBP performance. It was also found that the ideal bedding sand thickness ranges from 25mm to 50mm.

3.3.3 Gradation of Jointing Sand

Jointing sand is one of the basic materials used to construct ICBP. The role of jointing sand is inevitable in the design of ICBP. The primary function of jointing sand is to transfer load through shear action between one block to another and enhance the interlocking efficiency. Jointing sand plays a key role in the vertical and horizontal interlocking of concrete blocks. The lack of jointing sand between the blocks causes the blocks to act independently. Thus, load transfer will be inadequate, resulting in a negligible interlocking. The pavement without jointing sand deflects approximately 14% more than the pavement with jointing sand.[14] The jointing sand is laid between the IPB through brooming as the joint width is mostly less than 5mm. At times the jointing sand may come to the surface due to external agencies such as wind and water. The problem becomes adverse in the airports during the propulsion of aircraft. Polymer stabilisers are employed to prevent the removal of jointing sand during jet blasts of aircraft. It is also recommended that spacers be installed between the pavers to ensure uniform joint width.[42] However, the removal of jointing sand occurs only during the initial construction period and once interlocking begins, the jointing sand becomes stable. The maximum particle size is an important parameter influencing the function of the jointing sand. The size of the sand particles should be such that it should fill the gaps without voids and be strong enough to transfer the load. The range of particle size of jointing sand is considered as 1.2 mm and 1.8mm.[64,65] When the jointing sand has a particle size less than 1.0 mm, the overall deformation was less than 2mm after 10000 repetitions of load.[2] From the literature review, it was concluded that jointing sand plays a major role in the interlocking mechanism and load spreading ability of ICBP. Also, it was concluded that fine graded soil with an optimal joint width of 3–5mm is preferred over coarse graded soil. The different jointing sand gradations as per different literatures are provided in Table 3.3.

3.3.4 Joint Width

The function of the jointing sand depends on the joint width between the blocks and are interrelated to each other. The joints should be wide enough to allow the flow of jointing sand. It should also allow bedding sand intrusion while simultaneously being narrow enough to prevent excessive creep. It is determined that Zone III and Zone IV jointing sand should have joint widths of 4mm and 2.5mm, respectively.[14] The ideal joint width of 2–4mm should be provided for improved performance.[65] The wider joint width result in lower elastic moduli of the ICBP pavement section.[11] The joint width plays a major role in the interlocking mechanism of ICBP. The interlocking mechanism of ICBP accelerates with the increase in the traffic load and the formation of hinge which results in higher lateral load dissipation. The rotation of the blocks is the key factor for the hinge formation. The rotation of the block may end up with the failure of hinge formation, if the joint width is wider. The wider joint width may tend the block to move out from its position during the process of rotational interlocking. Therefore, the joint width for the construction of ICBP is limited to a maximum width of 2–4mm.

TABLE 3.3 Different jointing sand gradations

SIEVE SIZE (MM)	*PERCENTAGE PASSING (%)*				
	IRC SP 63[2]	*SHACKEL*[54]	*GARRY*[55]	*PANDA AND GHOSH*[14]	*ARJUN ET AL.*[56]
9.5	100	100	100	100	100
4.75	100	100	100	100	100
2.36	100	100	95–100	100	100
1.18	90–100	90–100	70–100	68.23	81.81
0.600	60–90	60–90	40–75	47.05	51.13
0.300	30–60	30–60	10–40	28.52	23.86
0.150	15–30	15–30	2–25	15.29	6.81
0.075	0–10	5–10	0–10	10	2.27

3.4 BASE AND SUBBASE LAYER

The base layer is an essential supporting layer for the ICBP. There are different types of base layers like granular base, cement-treated base, and bituminous treated base. The material conversion factor for cement bound material is 2–3mm, for Pavement Quality Concrete (PQC) it is 4, for Dense Bituminous Macadam it is 3, and for stabilised subgrade and subbase it is 0.5mm and 0.7mm. For example, a granular layer of 300mm can be replaced with a cementitious foundation layer of 100–150mm.[66] The significance of the ICBP wearing layer is greater for granular layer pavements than for cement or bituminous bound layers. Cement-bound bases are more resistant than granular materials and are also more cost-effective since the inferior quality of the aggregates can be improved by cement stabilisation.[67] The granular layer has a greater influence on surface deformation than the sand cushion layer or subgrade soil. When ICBP is laid in a granular layer, it has the least permanent deformation but has an elastic deflection of 1mm to 2mm. The vertical compressive strain above the subgrade layer is considered as the performance criteria.[2] Water penetration into the granular layer affects the overall strength of the granular layer. It is recommended to provide a drainage coefficient ranging from 0.4 to 1.4 based on the quality and duration of the drainage.[26]

The permanent deformation at the subgrade and shear failure at the joints with regard to the traffic load are the failure modes for ICBP when laid in granular material.[43] When a 200mm granular layer is used, the rut depth is measured as 15mm after 600,000 load repetitions. However, when the granular layer is extended to 300mm, the rut depth is determined to be 5.5mm after 1,000,000 repetitions of load.[12] This proves that the increase in the base thickness significantly reduces the rut depth of ICBP. From the literature review, it was found that granular material is the most commonly used material for the ICBP compared to cement and bituminous-treated base due to the cost efficiency of its granular layer. The base layer plays a vital role in the structural performance of ICBP.

3.5 SUBGRADE

The subgrade is the lowest supporting stratum of the ICBP section. The quality of the subgrade is a crucial aspect that determines the pavement deflection. California Bearing Ratio (CBR) is considered as the measure for the subgrade strength. The minimum CBR for the design of Interlocking Concrete Block Pavement is considered as 5%.[61] Few studies recommends the ICBP should be laid with a CBR above 10% and an elastic modulus greater than 80MPa.[67] When the subgrade CBR is weak, it is also recommended to utilise high-quality granular materials. The plate load tests, and field studies propose that the ICBP have a minimum CBR of 10% for low volume traffic and when the CBR is larger than 20%, no enhancement layer is required.[68] Rutting caused due to the vertical compressive strain above the subgrade is considered as one of the critical failures for the design of ICBP. The failure in the subgrade is reflected at the top of the ICBP wearing layer, resulting in surface unevenness. The placement of geotextiles on the subgrade increases the evenness of the ICBP wearing surface significantly. From the literature review, it was found that the subgrade plays a similar role in ICBP as in the case of flexible pavements. Rutting occurs at the top of the subgrade layer is considered as one of the critical parameters for the ICBP design. Therefore, the subgrade CBR plays a remarkable role in the performance of ICBP.

3.6 SUMMARY

The performance of ICBP is significantly influenced by the geometrical components of the wearing layer. The intensity of the interlocking mechanism of the IPB is directly proportional to the performance of the ICBP. The block shape, block size, block thickness, and block laying pattern have a major impact on the deflection and interlocking behaviour of ICBP. The deflection and interlocking behaviour of ICBP is moderately influenced by block strength and block laying angle. The edge restraint is regarded as a critical component that expedites the interlocking mechanism. The jointing and bedding sand are other parameters that affect the performance of the ICBP. The jointing sand should be stable and rigid in order to dissipate the applied stress laterally by shear action. The optimal joint width for the IPB must be maintained for better hinge formation. The supporting layers should be sufficiently stiff to decrease and sustain the applied stress. For the design of ICBP all the components and factors should be chosen effectively for the better performance of ICBP.

REFERENCES

1. Shackel, B. (1984) The Design of Interlocking Concrete Block Pavements for Road Traffic. *Second International Conference on Concrete Block Pavement,* Delft, The Netherlands.

2. Shackel, B. (1988) The Evolution and Application of Mechanistic Design Procedures for Concrete Block Pavements. *Third International Conference on Concrete Block Pavement*, Rome, Italy.
3. IRC SP 63 (2004) Guidelines for the use of Interlocking Concrete Block Pavement. *Indian Road Congress Special Publication*, New Delhi.
4. Livneh, M., I. Ishai, and S. Nesichi (1992) Development of a Pavement Design Methodology for Concrete Block Pavements in Israel. *Fourth International Conference on Concrete Block Pavement,* Auckland, New Zealand.
5. Knapton, J., and S.D. Barber (1979) The behaviour of a concrete block pavement. *Proceedings of the Institution of Civil Engineers,* 1(66), 277–292.
6. Van der Vlist, A.A, L.J.M. Houben, H.J. Dekkers, J.J. Duivenvoorden, J. van der Klooster, S.G. van der Kreef, and M. Leewis (1984) Behaviour of Two Concrete Block Test Pavements on a Poor Subgrade. *Second International Conference on Concrete Block Pavement,* Delft, The Netherlands.
7. Wellner, F., and T. Gleitz (1996) Dynamic Load Bearing Tests on Block Pavements. *Fifth International Conference on Concrete Block Pavement,* Tel Aviv, Israel.
8. Shackel, B. (1988) The Evolution and Application of Mechanistic Design Procedures for Concrete Block Pavements. *Third International Conference on Concrete Block Pavement*, Rome, Italy.
9. Panda, B.C., and A.K. Ghosh (2002) Structural behavior of concrete block paving. II: Sand in bed and joints. *Journal of Transportation Engineering*, 128(2), 130–135.
10. Mampearachchi, W.K., and W.P.H. Gunarathna (2010) Finite-element model approach to determine support conditions and effective layout for concrete block paving. *Journal of Materials in Civil Engineering*, 22(11), 1139–1147. https://doi.org/10.1061/(ASCE)MT.1943-5533.0000118
11. Ascher, D., T. Lerch, and F. Wellner (2006) Deformation Behaviour of Concrete Block Pavements Under Vertical and Horizontal Dynamic Load. *Eighth International Conference on Concrete Block Paving*, San Francisco, California USA.
12. Huurman, M., and W. Boomsma (2006) Mechanical Behaviour of a Permeable Base and Bedding Material and the Rutting Behaviour. *Eighth International Conference on Concrete Block Paving*. San Francisco, California USA.
13. Kuipers, G. (1992) The Choice of an Appropriate Block Shape for Heavy Industrial Flexible Pavements. *Fourth International Conference on Concrete Block Pavement*, Auckland, New Zealand.

14. Panda, B.C. (2006) Load Dispersion Ability of Concrete Block Layers. *Eighth International Conference on Concrete Block Paving*, San Francisco, California USA.
15. Mahapatra, G., and K. Kalita (2018) Effects of Interlocking and Supporting Conditions on Concrete Block Pavements. *Journal of The Institution of Engineers (India): Series A*, 99(1), 29–36.
16. Pham, N.P., W. Lin, D.G. Park, H. Kim, and Y.H. Cho (2014) Evaluation methodology for laying pattern of interlocking concrete block pavements using a displacement-moment concept. *Journal of Transportation Engineering*, 140(2), 04013008. https://doi.org/10.1061/(ASCE)TE.1943-5436.0000623
17. Lin, W., D. Kim, S. Ryu, H. Hao, Y.E. Ge, and Y.H. Cho (2018) Evaluation of the load dissipation behavior of concrete block pavements with various block shapes and construction patterns. *Journal of Materials in Civil Engineering.*, 30(2), 04017291. https://doi.org/10.1061/(ASCE)MT.1943-5533.0002113
18. Gunatilake, D., and W.K. Mampearachchi (2019) Finite element modelling approach to determine optimum dimensions for interlocking concrete blocks used for road paving. *Road Materials and Pavement Design*, 20(2), 280–296. https://doi.org/10.1080/14680629.2017.1385512
19. Eisenmann, J. and G. Leykauf (1988) Design of Concrete Block Pavements in FRG. *Third International Conference on Concrete Block Pavement*, Rome, Italy.
20. Dutruel, F., and J. Dardare (1984) Contribution to the Study of Structural Behaviour of a Concrete Block Pavement. *Second International Conference on Concrete Block Pavement,* Delft, The Netherlands.
21. Knapton, J., and S.D. Barber (1980) UK Research into Concrete Block Pavement Design. *First International conference on Concrete Block Paving*, Newcastle, England.
22. Miura, Y., M. Takaura, and T. Tsuda (1984) Structural Design of Concrete Block Pavements by CBR Method and its Evaluation. *Second International Conference on Concrete Block Pavement,* Delft, The Netherlands.
23. Festa, B., P. Giannattasio, and M. Pernetti (1996) Evaluation of Some Factors Influence on The Interlocking Paving System Performance. *Fifth International Conference on Concrete Block Pavement*, Tel Aviv, Israel.
24. Geller, R. (1996) Design Criteria of Concrete Block Pavements for Heavy and Industrial Traffic. *Fifth International Conference on Concrete Block Pavement,* Tel Aviv, Israel.

25. Raymond, S.R. (1984) Corps of Engineers Design Method for Concrete Block Pavements. *Second International Conference on Concrete Block Pavement,* Delft, The Netherlands.
26. Rada, G.R., R. David Smith, S. John Miller, and W. Matthew Witczak (1992) Structural Design of Concrete Block Pavements. *Third International Conference on Concrete Block Pavement*, Auckland, New Zealand.
27. Mascio, P., L. Moretti, and A. Capannolo (2019) Concrete block pavements in urban and local roads: analysis of stress-strain condition and proposal for a catalogue. *Journal of Traffic and Transportation Engineering (English Edition)*, 6(6), 557–566. https://doi.org/10.1016/j.jtte.2018.06.003
28. Muraleedharan, T., and V. K Sood (2003) Past and Present Efforts for Popularisation of Interlocking Concrete Block Pavement Technology in India. *International Conference on Concrete Block Paving (PAVE AFRICA 2003)*, 3–11.
29. Soutsos, M.N., K. Tang, H.A. Khalid, and S. G. Millard (2011) The effect of construction pattern and unit interlock on the structural behaviour of block pavements. *Construction and Building Materials*, 25(10), 3832–3840. https://doi.org/10.1016/j.conbuildmat.2011.04.002
30. Knapton, J. (1976) The Design of Concrete Block Roads, Report No. 42.515. *Cement and Concrete Association, Wexham Springs*, United Kingdom.
31. Mampearachchi, W. K., and A. Senadeera (2014) Determination of the most effective cement concrete block laying pattern and shape for road pavement based on field performance. *Journal of Materials in Civil Engineering*, 26(2), 226–232. https://doi.org/10.1061/(ASCE)MT.1943-5533.0000801
32. Yaginuma, H., I. Takashi, and I. Takuya (1998) Evaluation on Durability of Interlocking Block Pavement Under Repeated Loading by Heavy Vehicles. *Third International Workshop on Concrete Block Paving*, Cartagena de Indias, Colombia.
33. Nishizawa, T. (2003) A Tool for Structural Analysis of Block Pavements Based on 3DFEM. *Seventh International Conference on Concrete Block Pavement Sun City*, South Africa.
34. PA01 (2014) Concrete Segmental Pavements – Detailing Guide. *Concrete Masonry Association of Australia*, New South Wales, Australia.
35. Algin, H.M. (2007) Interlock mechanism of concrete block pavements. *Journal of Transportation Engineering,* 133(5), 318–326. https://doi.org/10.1061/(ASCE)0733-947X(2007)133:5(318)

36. IRC 58 (2015) Guidelines for Plain Jointed Rigid Pavements for Highways Pavements. *Indian Road Congress*, New Delhi.
37. IS 15658 (2006) Precast Concrete Blocks for Paving – Specification. *Bureau of Indian Standards,* New Delhi.
38. Alex Visser, T. (2006) Deterioration of Concrete Block Pavements. *Eight International Conference on Concrete Block Paving*, San Francisco, California.
39. Houben, L.J.M., A.A.A. Molenaar, G.H.A.M. Fuchs, and H.O. Moll (1984) Analysis and Design of Concrete Block Pavements. *2nd International Conference on Concrete Block Paving*, Delft, The Netherlands, 86–97.
40. Filho, J.N.S., S.N. da Silva, G.C. Silva, J.C. Mendes and R.A.F. Peixoto (2017) Technical and environmental feasibility of interlocking concrete pavers with iron ore tailings from Tailings Dams. *Journal of Materials in Civil Engineering*, 29(9), 04017104. https://doi.org/10.1061/(ASCE)MT.1943-5533.0001937
41. Allan Dowson, J. (1992) Fact or Fiction A Review of Test Methods for Concrete Small Element Paving. *Fourth International Conference on Concrete Block Pavement*, Auckland, New Zealand.
42. Emery, A.J. (1988) An Evaluation of the Performance of Concrete Blocks on Aircraft Pavements at Luton Airport U.K. *Third International Conference on Concrete Block Pavement*, Rome, Italy.
43. Sun, A.L. (1996) Design Theory and Method of Interlocking Concrete Block Paving for Port Areas. *Fifth International Conference on Concrete Block Pavement*, Tel Aviv, Israel.
44. Barber, S.D., and J. Knapton (1980). An experimental investigation of the behaviour of a concrete block pavement with a sand sub-base. *Proceedings of the Institution of Civil Engineers*, 69(1), 139–155.
45. Sharp, K.G., and Simmons, M.J. (1980). Interlocking Concrete Blocks: State of the Art Review. *Australian Road Research Board (ARRB) Conference,* Vol. 10, No. 2, Sydney.
46. Mujaj, L., and Smith, D.R. (2001). Evolution of interlocking concrete pavements for airfields. *Advancing Airfield Pavements, Proceedings of the 2001 Airfield Pavement Specialty Conference*, ASCE, Chicago, Illinois, pp. 253–266.
47. Frank Bullen (1996) Edge Restraints for Segmental Concrete Block Pavements. *Fifth International Conference on Concrete Block Pavement*, Tel Aviv, Israel.
48. Hodgkinson, J., and C.F. Morrish (1980) An Interim Guide to the Design of Interlocking Concrete Pavements. *Cement and Concrete Association of Australia*, New South Wales, Australia, Publication No TN 34.

49. Knapton, J., and H.M. Algin (1998). Research into the structural performance of interlocking block pavements. *Third International Workshop on Concrete Block Paving*, Colombia.
50. Panda, B.C., and A.K. Ghosh (2002) Structural behavior of concrete block paving. I: Concrete blocks. *Journal of Transportation Engineering*, 128(2), 123–129. doi:10.1061/(ASCE)0733-947X(2002)128:2(123).
51. Knapton, J., and I.D. Cook (2000) Total quality design of pavements surfaced with pavers. *Journal of Transportation Engineering*, 126(3), 249–256. https://doi.org/10.1061/(ASCE)0733-947X(2000)126:3(249)
52. IS 383 (2016) Coarse and Fine Aggregate for Concrete – Specification. *Bureau of Indian Standards*, New Delhi.
53. ASTM Standard C33 (2003) Specification for Concrete Aggregates. *ASTM International,* West Conshohocken, PA.
54. Shackel, B. (1980) An Experimental Investigation of the Roles of the Bedding and Jointing Sands in the Performance of Interlocking Concrete Block Pavements *Concrete/Beton*, No. 19, Newcastle, England.
55. Anderton, Gary I. (1992) Military Applications of Block Pavements in the United States *Fourth International Conference on Concrete Block Pavement,* Auckland, New Zealand.
56. Arjun Siva Rathan, R.T., V. Sunitha., P. Murshidha., and V. Anusudha (2020) Influence of bedding and jointing sand on the shear strength characteristics of Interlocking Paver Blocks – Bedding sand interface, *International Journal of Pavement Engineering,* 23(7), 2160–2175. https://doi.org/10.1080/10298436.2020.1847286
57. Knapton, J., and M. O'Grady (1983) Structural behavior of concrete block paving. *Journal of Concrete Society*, 17, 17–18.
58. Miura, Y., and Katsuhiko. M (1988) Effects of Geotextiles on Development of Rutting of Concrete Block Pavement under Repeated. *Third International Conference on Concrete Block Pavement*, Rome, Italy.
59. ASCE 58-16 (2016) Structural Design of Interlocking Concrete Pavement for Municipal Streets and Roadways. *American Society of Civil Engineers*, Virginia, United States.
60. Mavin, K.C. (1980) Interlocking Block Paving in Australian Residential Streets. *First International Conference on Concrete Block Pavement*, Newcastle, England.
61. IRC SP 63 (2018) Guidelines for the use of Interlocking Concrete Block Pavement. *Indian Road Congress Special Publication*, New Delhi.
62. Simmons, M.J. (1979) Construction of interlocking concrete block pavements. *Proceedings of Australian Road Research,* 90, 71–80.

63. Arjun Siva Rathan, R.T., and V. Sunitha (2021) Development of deflection prediction model for the interlocking concrete block pavements. *Transportation Research Record: Journal of the Transportation Research Board*, 2676(3), 292–314. https://doi.org/10.1177/03611981211051339
64. Aly, M.A., S. Wonosaputra, and P. Iskandar (1986) The First Interblock Paving Rural (Intercity) Road Section in Indonesia. *International Workshop on Interlocking Concrete Pavements,* 1986, Melbourne, Australia.
65. Ryntathiang, T.L., M. Mazumdar, and B.B. Pandey (2006) Concrete Block Pavement for Low Volume Roads. *Eighth international conference on concrete block paving*, San Francisco, CA, USA.
66. Cook, I. D., and J. A. Knapton (1996) Design Method for Lightly Trafficked and Pedestrian Pavements. *Fifth International Conference on Concrete Block Pavement*, Tel Aviv, Israel.
67. Judycki, J., J. Alenowicz, and W. Cyske (1996) Structural Design of Concrete Block Pavement Structures for Polish Conditions. *Fifth international conference on concrete block paving*, Tel Aviv, Israel.
68. Mampearachchi, W. K., and W. P. H. Gunarathna (2010) Finite-element model approach to determine support conditions and effective layout for concrete block paving. *Journal of Materials in Civil Engineering*, 22(11), 1139–1147. https://doi.org/10.1061/(ASCE)MT.1943-5533.0000118

Structural Analysis of ICBP 4

4.1 INTRODUCTION

The lateral load dissipation of the ICBP wearing layer through the interlocking mechanism results in a unique stress distribution in the ICBP. It is crucial to comprehend the various structural characteristics of ICBP. The performance of the ICBP is greatly influenced by structural characteristics such as rutting, deflection, stress, and shear stress. Deflection is regarded as the important metric to be assessed in order to evaluate the performance of the ICBP. The plate load test is widely used to determine the deflection behaviour of ICBP. Another structural characteristic that affects the ICBP's performance is stress and strain. Due to the interlocking mechanism of the ICBP, stress is dispersed laterally. The stress distribution on each layer is measured using pressure cells. Another crucial characteristic of the ICBP is its resistance to shear stress, which is more likely to develop during vehicle acceleration and braking at intersections. The ICBP is highly susceptible to shear stress because of its distinct wearing layer. Shear stress is calculated using large-scale direct shear. With the exception of the wearing layer, ICBP performs identically to flexible pavements. Fatigue and rutting are the two failure criteria for flexible pavement design. Due to the distinct wearing layer, fatigue failure is disregarded in the design of the ICBP. Rutting of ICBP occurs as a result of vertical strain above the subgrade as well as vertical deformation of the bedding sand. The accelerated pavement testing machine is used to evaluate rutting due to subgrade failure, and the large-scale direct shear test is used to determine rutting in the bedding sand layer.

4.2 INTERLOCKING MECHANISM

The ICBP has a discrete wearing layer that is distinct from the continuous wearing layer of flexible and rigid pavement. Similarly, the load distribution

DOI: 10.1201/9781003432371-4

behaviour of the ICBP differs from that of the flexible pavement. Flexible pavement distributes loads through grain-to-grain transfer, whereas rigid pavement is dominated by slab action. The load dispersion of the ICBP is accomplished through shear transfer through an interlocking mechanism. The resistance to horizontal movement between blocks is referred to as interlocking. Due to lipping and wedging action, the load is transferred between the blocks via frictional transfer through jointing sand.[1] As a result, the interlocking mechanism of ICBP is critical to the load distribution behaviour of ICBP.

The interlocking mechanism is accomplished through three different methods: translational interlocking, rotational interlocking, and vertical interlocking. Translational interlocking is a type of horizontal interlocking. The horizontal force is initiated during the breaking and acceleration of vehicles. This horizontal force tends to move the block horizontally. But the presence of neighbor blocks and the jointing sand between the blocks restrict the horizontal movement through restricted force. This results in the horizontal interlocking as shown in Figure 4.1. When the horizontal force is higher than the restricted force, the block may tend to rotate but is unable to come out independently as the rotation of the block is restricted by the wedge of the nearer block. This interlocking that occurs to the resistive force against rotation is called rotational interlocking. Vertical interlocking is primarily actuated by the shear mechanism and the stiffness of the jointing sand. The presence of jointing sand facilitates load transfer by spreading the load wider, depending on the interlocking efficiency.[2,3,4]

The jointing sand is identified as a primordial media in the ICBP interlocking mechanism. Gradations of bedding and jointing sand facilitate the transfer of shear forces and enhance interlocking mechanisms. Aside from the wearing layer, supporting layers such as granular layer stiffness and subgrade California Bearing Ratio (CBR) play a key role in ICBP performance. The jointing sand is found to be a primitive medium in the interlocking mechanism in ICBP. In addition to jointing sand, the wearing layer of ICBP further aids the interlocking. Gradations in bedding and jointing sand aid in the shear transfer and improve interlocking mechanisms. Apart from the wearing layer, supporting layers such as the rigidity of the granular layer and subgrade California Bearing Ratio (CBR) also play an important role in the performance of ICBP. The interlocking mechanism also enhances the ICBP wearing layer's elastic modulus.[5]

There is no direct method for evaluating the interlocking mechanism of ICBP. The deflection and stress curve obtained from the plate load test is an indirect method for understanding the interlocking mechanism. The

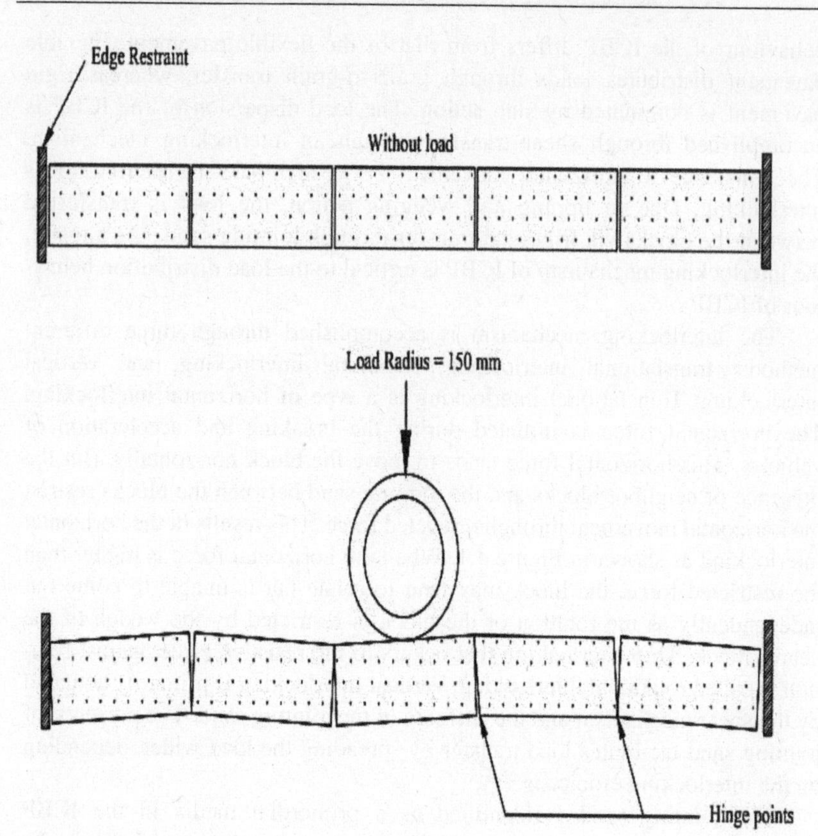

FIGURE 4.1 Interlocking mechanism of ICBP.

D-Moment method is one of the experimental assessment alternatives for estimating the interlocking mechanism of ICBP. A horizontal load of 20mm/min is applied to the blocks placed on the surface of the frame to evaluate the interlocking mechanism. The response of the blocks to the application of horizontal load is recorded and analysed. The D-moment parameter indicates the rotations and shifts of individual blocks and predicts their behaviour. As indicated in Equation 1, the interlocking mechanism of the entire blocks is specified by the sum of the movement and rotation of each individual block.[3]

$$D-moment = \sum_{i,j=1}^{n} \left| Rij \times dij \right|^{3} \tag{4.1}$$

Where, Rij = displacement from the original position to a temporary position at time tk; dij = distance from the moving point Ai to the line of centre point movement; and nij = number of blocks

Block rotation is measured in terms of di, and block movement is represented by Ri. Three components constitute the interlocking mechanism during the horizontal load application. The initial phase represents the linear increase in deflection with applied load. Blocks reach their maximum rigidity during the first phase, which is referred to as the stiffness phase. The second phase reflects the length of interlocking achieved by the transfer of load between blocks through jointing sand with the same or decreased interlocking.

4.3 DEFLECTION BEHAVIOUR

Deflection is one of the essential structural parameters used to analyse ICBP behaviour. The deflection of the flexible pavement is limited to 0.5mm to avoid the fatigue cracking due to the traffic load. Since the ICBP has a distinct layer, fatigue cracking problems are not encountered. Consequently, when compared to flexible pavement, the ICBP has a higher allowable deflection. The deflection of ICBP is impacted by the presence of bedding sand and the initial movement of blocks.[6,7] The deflection of ICBP is impacted by the presence of bedding sand and the initial movement of blocks. Typically, the ICBP exhibits elastic deflection with a maximum allowable limit of 4mm, while yielding relatively minor rutting deformation.[8] The ratio of the unloaded to the loaded deflection at the joints, as measured by a falling weight deflectometer (FWD), is used to calculate the load transfer efficiency.

The deflection is one of the failure criteria for the design of the ICBP. The empirical design of the ICBP assumed that the ICBP would fail owing to greater deflection without cracking the Interlocking paver Block (IPB).[10] Under the same applied load, the deflection of the base layer is higher than that of the subgrade layer. This is because the base layer absorbs and reduces the applied load prior to transmitting it to the subgrade layer.[11] The majority of studies used plate load testing to determine the deflection behaviour of ICBP.[12,13] Light Weight Deflectometer (LWD) is employed in assessments to quantify the deflection using plate load tests with dynamic loading.[12] Falling Weight Deflectometer (FWD) and Benkelman Beam Deflectometer Method (BBD) are two extensively deployed pieces of equipment for determining the deflection of ICBP in field research.

There are empirical methods to determine the deflection of ICBP without experimental testing.[14] The deflection can be predicted from a deflection prediction model developed from plate load testing and numerical analysis. The parameters in the equations are provided based on the deflection test results incorporating different geometric parameters. The Equation 4.2[14] is used to determine the deflection of ICBP with zigzag block shape laid in herringbone pattern. Similarly, the deflection of ICBP laid in zigzag block shape in stretcher pattern is determined using Equation 4.3.[14] The correction factor C is applied when the applied load is greater than 40 kN. The deflection of the ICBP can be determined for different loading and supporting conditions.

$$\text{Deflection} = (0.787 - 0.089SF - 0.001E_{Gran} - 0.004\,E_{SG} + 1.086P) \times C \tag{4.2}$$

where, SF – Shape factor, E_{Gran} – Elastic modulus of base layer, E_{SG} – Elastic modulus of subgrade. C- correction factor, C=1 for load up to 40kN and C = 0.98 for load greater than 40kN.

$$\text{Deflection} = (0.842 - 0.089SF - 0.002E_{Gran} - 0.004E_{SG} + 1.335P) \times C \tag{4.3}$$

where, $E_{Gran} = 0.2*h^{0.45}*E_{SG}$, $E_{SG} = 10 * \text{CBR}$ for CBR≤ 5, E_{SG} = 17.6 * (CBR)0.64for CBR > 5 %, h = Thickness of granular layer (mm), CBR = California Bearing Ratio (%)

The pavement deflection curve derived from different plate load tests reveals a nonlinear curve.[15] It is interesting to note that the rate of deflection reduces as the applied force increases, and that vertical deflection is three times that of horizontal deflection.[11,12] This is due to the fact that an increase in the loading rate enhances the rigidity of the pavement and the interlocking of the blocks which is quite different from flexible and rigid pavement principles. Despite the fact that the trend of the deflection is non-linear, the initial portion of the load-deflection curve is linear, and the slope varies as the load increases. This is an indirect demonstration on the interlocking effect of ICBP. The interlocking mechanism is further well proven from static repetitive plate load test and FWD test.[16] The static repetitive plate load test also reveals the elastic properties of the ICBP in relation to the number of cycles. The initial phase of the loading and unloading cycle is attributed to an increased degree of deflection. This is due to the initial energy loss during bedding sand compaction. The deflection and recovery are found to be optimum after 150–200 loading cycles, indicating that the pavement gained a fully elastic property, establishing that block pavements stiffen progressively as the number of load repetitions increases.[15]

4.4 VERTICAL STRESS BEHAVIOUR

Stresses and strains are two basic structural factors that must be considered while designing a pavement. In the case of flexible pavements, the vertical compressive strain and horizontal tensile strain are the two critical failure criteria of flexible pavements. In the case of the ICBP, the stress distribution or load transfer mechanism is unique and depends on several geometric features of the ICBP wearing layer. The magnitude of the applied stress is considerably reduced in the low-lying layers of ICBP, resulting in lower total pavement thickness compared to flexible pavement.[17] The individual blocks under the traffic load are more efficient in transferring the load laterally. Due to this increased ability to lateral distribution of load, the stress accumulated at the top of the base is reduced to 35% of the applied stress.[18] Stress transferred from the surface to the sublayers decreases when the interlocking effect between individual blocks increases.[12]

The vertical stress and its distribution are evaluated using pressure cells positioned beneath the applied load in the plate load test. The applied stress is considerably reduced below the IPB. This is because the applied load develops a lateral distribution of load and reduces the applied stress significantly. The magnitude of lateral load dissipation is dependent on the geometry parameters of the IPB, including block shape, laying pattern, and block thickness. The elastic modulus of the IPB plays a significant impact in reducing the applied stress at the top of the base layer. The effectiveness of the interlocking is improved by an increase in elastic modulus. Because of their small size, individual concrete blocks experience compressive stress with negligible bending stress.[13] Similarly, the thickness of the IPB plays a major role in the stress reduction of IPB. The observed stress beneath the block is greater at 60mm block thickness and decreases as the block thickness increases.[19] The reduction in the block thickness is because of the increase in the vertical surface area which considerably increases the lateral load dissipation through an interlocking mechanism. However, further increase in block thickness demonstrates only a small reduction in the measured stress below the block.

Bedding sand provided below the concrete blocks plays a negligible role in the stress absorption of the applied stress. The development of stress in paver bedding sand is complex and depends on the geometry properties of the paver.[1] This is due to the lower thickness of the bedding sand. Moreover, the modulus of the IPB and the base layers are higher and act as a rigid medium between the bedding layer. Therefore, the bedding sand is subjected to higher deflection rather than sustaining the applied stress. Beneath the bedding sand, a granular or cement-bound subbase is provided to reduce stress on the subgrade. Vertical stress on the granular base are 10 to 30 percent less than those on the cement stabilised base.[20] Vertical stresses are reduced by 30–40% when the subgrade modulus is improved from 60 to 100 MPa.

The stress distribution is proportional to the interlocking effect of the ICBP. The stress at the lower layers are decreased with the increase in the load repetitions. The stress at the interface between the subgrade and base layer is vital to reduce the vertical compressive strain above subgrade which leads to rutting failure. The repetition of load enhances the interlocking between blocks and reduces the stress accumulated in the lower layers. This is another merit in the ICBP that the rutting caused due to the subgrade deflection is lower when compared to conventional flexible pavement. This reduction in rutting also aids in the thickness of the supporting layer and thereby reduces the total pavement thickness. Since different factors possesses significant impact on the vertical stress behaviour of ICBP, the stress based ICBP design are contradictory and not recommended.[1]

4.5 SHEAR STRESS BEHAVIOUR

The shear strength of ICBP is one of the factors which influences the serviceability and performance of ICBP. ICBP is most affected by shear stress while the vehicle is subjected to braking, accelerating, or turning. The ratio of shear stress/shear strength of ICBP was originally developed for airfield pavements where the shear stresses from aircraft loads and tyre pressures are high relative to the strengths of the subgrade materials.[21] A similar concept is applied to evaluate the shear stress behaviour of ICBP. The shear strength of the ICBP is evaluated by the conventional large scale direct shear apparatus. The test procedure was carried out as per ASTM D5321/D5321M (2019). The test set-up consists of two square boxes, each of size 300 mm × 300 mm × 100 mm. The lower box was mounted on a bearing frame, which ensures a lateral shearing to the load. The shear rate for all the testing was 1.25mm/min. The lower portion of the box was filled with the prepared bedding sand for a required weight and compacted. The compacted soil should ensure 95% relative compaction. IPB occupied the upper portion of the box with the joint width filled by jointing sand. The horizontal displacement, vertical displacement and load were recorded by two LVDTs along with a load cell attached to a data logger.

The shear stress to the shear strength ratio provides an indirect measure of the rutting in the bedding sand and IPB interface. The rutting due to bedding sand is evaluated as the ratio of shear stress to the Shear Strength (SSR). The shear strength of the IPB – bedding sand interface is determined from large scale direct shear test as shown in Equation 4.4 and 4.5.[22] The different categories aligned to the level of rutting risk were defined as follows:[22]

SSR < 0.3, low risk of rutting
0.3 ≤ SSR ≤ 0.7, medium risk of rutting
SSR > 0.7, high risk of rutting

$$Shear\ Strength\ Ratio(SSR) = \frac{\tau_f}{\tau_{max}} \tag{4.4}$$

$$\tau_{max} = c + \sigma tan\varphi \tag{4.5}$$

Where, τ_f= Applied Shear Stress, τ_{max}= Shear strength of the material, c = cohesion, Ø = Angle of internal friction, σ = Normal stress

Another method for determining shear strength has been designed exclusively for the ICBP to analyse vertical and horizontal shear stress.[23] Three blocks are lined up in a row on a multilayered piece of wood in the test setup for determining vertical stress. The joints with 5mm are filled with precious-crushed sand and sideways closed with a piece of soft wood. Using a mallet, the sand is mechanically compacted through vibration. After applying a constant horizontal load, the centre block is pushed out with a gradually increasing vertical force, and the vertical motion is recorded by a linear variable differential transformer located in the centre (LVDT).[23] The shear stress in the vertical joints can be calculated using the Equations 4.6 and 4.7.[23]

$$\sigma_n = \frac{H - R}{bh} \tag{4.6}$$

$$\tau_{nt} = \frac{V - G_s}{bh} \tag{4.7}$$

Where H = constant horizontal load, V = Continous Vertical load, R = Slipping friction force between the concrete block and wooden panel, G_s= dead load of paver block.

A similar setup used in the evaluation of vertical shear is used to determine horizontal joint behaviour. The test setup is comprised of three blocks arranged in a row on a multilayer wooden board. The sideways joints are sealed with crushed sand. Under dry conditions, the sand was mechanically compacted. In this test, a continuous horizontal force is measured while removing the middle block after a constant horizontal load is applied.[23] The shear stress in the horizontal joints can be calculated using the formula shown in Equation 4.8 and 4.9.[23]

$$\sigma_n = \frac{H - 1.5R}{bh} \tag{4.8}$$

$$\tau_{nt} = \frac{S - R}{2bh} \tag{4.9}$$

Where H = constant horizontal load, S = Continuous Horizontal load, R = Slipping friction force between the concrete block and wooden panel.

4.6 RUTTING BEHAVIOUR

Rutting is a vertical depression that occurs along the wheel path and is reported as a crucial failure criterion for ICBP.[24] The test pavement with ICBP experience less than 3mm rutting even after 11000 repetitions of 40kN. Rutting occurs faster for the first 2000 repetitions, then increases by 1.5mm to 2mm over the subsequent 9000 repetitions. After 100000 repetitions, rutting would be approximately 15 to 20mm.[25] The permissible rut depth for low volume roads is 25mm to 50mm.[26] For traffic loads larger than 2 MSA, the maximum rut depth limit considered is 20mm.[27] The rutting on the ICBP test track laid was less than 20mm.[28] The acceptable rutting values for regular roads with heavy traffic in Japan are specified to be between 30 and 40 mm.[28] In the Netherlands, a terminal rut depth of 25 mm is recommended as the limit of serviceability in a block pavement, whilst a rut depth of 35 mm is set as the structural limit.[29] The rutting behaviour of ICBP depends on the degree of compaction of the base layer and also the subgrade rutting is lower in stable pavements with superior interlocking, ensuring a long design life.[24] Due to the interlocking mechanism of the ICBP wearing layer, the rutting caused by the vertical compressive strain in the subgrade is lower. This is considered as one of the unique merits of the ICBP.[30]

The rutting of ICBP can be evaluated using the accelerated pavement testing, FWD and Heavy Vehicle Simulator (HVS). The accelerated pavement testing is one of the popular and widely used methods among the different evaluation methods. The accelerated pavement testing is carried out with a setup consisting of a dual wheel set with an axle wheel load of 60 to 80 kN. The wheel is allowed to move to and from and the number of cycles is recorded.[25] The measurement of the deflection and rutting is taken at frequent cycle intervals. The magnitude of the rut depth decreased with the number of repetitions for the ICBP. There are empirical equations to calculate the rut depth of the ICBP. The permanent deformation of ICBP with granular base is estimated from the Equation 4.10:[31]

$$D_R = 1.9 \times C \times l \times e^{-11h} \times N_m^{0.265} \tag{4.10}$$

Where D_R= Rut Depth; h = thickness of block (cm); l= Deflection of pavement under vehicle (mm); N_m = Cumulative number of axle load related to loading; C = 1.17, 1.25, 1.34 respectively for h = 8cm, 10cm, 12cm respectively.

The rut depth equation is developed from the test results derived from the plate load test and accelerated pavement testing.[12] The rut depth equation is the modified calculation of the rut depth determined in Equation 4.11.[12]

$$R = 961 \times C \times D_0 \times P_0^{-0.16} \times r_0^{-2.21} \times e^{(-0.24D_0 - 0.11h)} \times N^{0.265} \tag{4.11}$$

Where D_R= Rut Depth (mm); h = thickness of block (cm); D_0= resilient Deflection (mm); N_m = Cumulative number of axle load related to loading; P_0 = load pressure (MPa); C = 1.17, 1.25, 1.34 respectively for h = 8cm, 10cm, 12cm respectively.

4.7 SUMMARY

The deflection is one of the main failure criteria for the ICBP due to the presence of a discrete wearing layer. The permissible limit of the deflection value is higher than conventional flexible pavement due to the initial compression of the bedding sand. The stresses on different layers are measured using pressure cells. The stress distribution of ICBP depends on the geometric properties of IPB and the magnitude of interlocking. The shear stress is another parameter which influences the interface friction behaviour of ICBP. The rutting plays a significant role in the behaviour of ICBP and is considered as a crucial parameter for the design of ICBP. Overall, the different behaviour of ICBP mainly depends on the interlocking effect of the ICBP, which in turn relies on the geometric properties of the IPB.

REFERENCES

1. Algin, H.M. (2007) Interlock mechanism of concrete block pavements. *Journal of Transportation Engineering,* 133(5), 318–326. https://doi.org/10.1061/(ASCE)0733-947X(2007)133:5(318)
2. Mampearachchi, W.K., and A. Senadeera (2014) Determination of the most effective cement concrete block laying pattern and shape for road pavement based on field performance. *Journal of Materials in Civil Engineering*, 26(2), 226–232. https://doi.org/10.1061/(ASCE)MT.1943-5533.0000801
3. Pham, N.P., W. Lin, D.G. Park, H. Kim, and Y.H. Cho (2014) Evaluation methodology for laying pattern of interlocking concrete block pavements using a displacement-moment concept. *Journal of Transportation Engineering*, 140(2), 04013008. https://doi.org/10.1061/(ASCE)TE.1943-5436.0000623
4. Mampearachchi, W. (2019) Handbook on Concrete Block Paving. *Springer*, Singapore.
5. Jamshidi, A., K. Kurumisawa, G. White, T. Nishizawa, T. Igarashi, T. Nawa, and J. Mao (2019) State-of-the-art of interlocking concrete block pavement technology in Japan as a post-modern pavement. *Construction and Building Materials*, 200, 713–755.
6. Shackel, B. (1986) A Review of Research into Concrete Segmental Pavers in Australia Workshop on Interlocking Concrete Pavements. *Workshop on Interlocking Concrete Pavements*, Melbourne, Australia.
7. Miura, Y., M. Takaura, and T. Tsuda (1984) Structural Design of Concrete Block Pavements by CBR Method and its Evaluation. *Second International Conference on Concrete Block Pavement,* Delft, The Netherlands.
8. Shackel, B. (1978) An experimental investigation of the response of interlocking block pavements to simulated traffic loading, *Australian Road Research Board, Research Report (ARR)*, 90, 11–44.
9. Khanal, S., S.L Tighe, and R. Bowers (2013) Pavement performance mechanics of interlocking concrete paver crosswalk designs. *Canadian Journal of Civil Engineering*, 40(7), 583–594.
10. Soutsos, M.N., K. Tang, H.A. Khalid, and S.G. Millard (2011) The effect of construction pattern and unit interlock on the structural behaviour of block pavements. *Construction and Building Materials*, 25(10), 3832–3840. https://doi.org/10.1016/j.conbuildmat.2011.04.002
11. Sarkar, H., C. Halder, and T.L. Ryntathiang (2014) Behavior of interlocking concrete block pavement over stone dust grouted subbase.

International Journal of Advanced Structures and Geotechnical Engineering, 3(1), 3832–3840.

12. Lin, W., Y.H. Cho, and I.T. Kim (2016) Development of deflection prediction model for concrete block pavement considering the block shapes and construction patterns. *Advances in Materials Science and Engineering*, 2016. https://doi.org/10.1155/2016/5126436
13. Panda, B.C., and A.K. Ghosh (2002a) Structural behavior of concrete block paving. I: Concrete blocks. *Journal of Transportation Engineering*, 128(2), 123–129. doi:10.1061/(ASCE)0733-947X(2002)128:2(123)
14. Arjun Siva Rathan, R.T., and V. Sunitha (2021) Development of deflection prediction model for the interlocking concrete block pavements. *Transportation Research Record: Journal of the Transportation Research Board*, 2676(3), 292–314. https://doi.org/10.1177/03611981211051339.
15. Panda, B.C., and A.K. Ghosh (2002b) Structural behavior of concrete block paving. II: Sand in bed and joints. *Journal of Transportation Engineering*, 128(2), 130–135.
16. Paulo Roberto da Silva Morgado (2008) Design of block pavements. *M. Tech Dissertation*, Instituto Superior Técnico, Portugal.
17. Raymond, S.R. (1984) Corps of Engineers Design Method for Concrete Block Pavements. *Second International Conference on Concrete Block Pavement,* Delft, The Netherlands.
18. Clark, A.J. (1978) Block paving research and development. *Cement and Concrete Association*, United Kingdom, 12(7), 24–25.
19. Shackel, B. (1979) An Experimental Investigation of the Response of Interlocking Concrete Block Pavements to Simulated Traffic Loading. *Australian Road Research Board Conference Proc* (No. 90), Melbourne, Australia.
20. Arjun Siva Rathan, R.T., V. Sunitha, and V. Anusudha (2021) Parametric study to investigate the deflection and stress behaviour of interlocking concrete block pavement. *Road Materials and Pavement Design,* 23(10), 2293–2316. https://doi.org/10.1080/14680629.2021.1963819
21. Thompson, M., F. Gomez-Ramirez, and M. Bejarano (2002) Illi-Pave Based Flexible Pavement Design Concepts for Multiple Wheel Heavy Gear Load Aircraft. Proceedings 9th International Conference on Asphalt Pavements. Copenhagen, *International Society of Asphalt Pavements*, Denmark.
22. Jones, D.J., H. Li, R. Wu, J.T. Harvey, and D.R. Smith (2017) Full-scale structural testing of permeable interlocking concrete pavement to develop design guidelines. *International Low Impact Development Conference 2016*, Portland, Maine, 143–154. https://doi.org/10.1061/9780784480540.017

23. Füssl, J., W. Kluger-Eigl, and R. Blab (2016) Experimental identification and mechanical interpretation of the interaction behaviour between concrete paving blocks. *International Journal of Pavement Engineering*, 17(6), 478–488.
24. Huurman, M., and W. Boomsma (2006) Mechanical Behaviour of a Permeable Base and Bedding Material and the Rutting Behaviour. *Eighth International Conference on Concrete Block Paving,* San Francisco, California USA.
25. Ryntathiang, T.L., M. Mazumdar and B.B. Pandey (2006) Concrete Block Pavement for Low Volume Roads. *Eighth International Conference on Concrete Block Paving*, San Francisco, CA, USA.
26. AASHTO (1993) Guide for Design of Pavement Structures. *American Association of State Highway and Transportation Officials,* United States.
27. IRC 37 (2018) Guidelines for the Design of Flexible Pavements. *Indian Road Congress*, New Delhi.
28. Miura, Y., M. Takaura, and T. Tsuda (1984) Structural Design of Concrete Block Pavements by CBR Method and its Evaluation. *Second International Conference on Concrete Block Pavement,* Delft, The Netherlands.
29. Houben, L.J.M., A.A.A. Molenaar, G.H.A.M. Fuchs and H.O. Moll (1984) Analysis and Design of Concrete Block Pavements. *2nd International Conference on Concrete Block Paving*, pp. 86–97.
30. Barber, S.D., and Knapton, J. (1980) An experimental investigation of the behaviour of a concrete block pavement with a sand sub-base. *Proceedings of the Institution of Civil Engineers*, 69(1), 139–155.
31. Sun, A.L. (1996) Design Theory and Method of Interlocking Concrete Block Paving for Port Areas. *Fifth International Conference on Concrete Block Pavement*, Tel Aviv, Israel.

Structural Design of ICBP

5

5.1 INTRODUCTION

The distinct wearing layer contributes to the complexity of the structural design of the ICBP. Each nation has its own design methodologies and standards. The majority of nations developed ICBP designs based on flexible pavement design principles, presuming that the wearing layer is the only variance between the ICBP and flexible pavement in terms of performance. Like those in flexible pavement, failures in the supporting layers will be reflected in the wearing layer of ICBP. However, the transmission of applied load to the supporting layer of ICBP differs from that of flexible pavement. The design of ICBP for road pavements are determined based on design catalogues generated on experience, empirical method, and mechanistic-empirical design method. Similarly, the port and airport pavements are subjected to a heavy load and hence requires a specified design methodology. The airport and port design follows conventional flexible pavement design methodology and provides equivalence factor to adopt ICBP wearing layer thickness. The design software aids in determining the ICBP pavement thickness based on the traffic data and the geometric properties of Interlocking Paver Block (IPB).

5.2 DESIGN OF ICBP – LIGHT AND HEAVY TRAFFIC PAVEMENT

5.2.1 Design Based on Catalogue

Based on experience, the catalogue for the design of ICBP section is developed. The catalogue aids in attaining the thickness based on fixed traffic condition

DOI: 10.1201/9781003432371-5

and soil type. One such catalogue is developed by Indian Road Congress[1] (IRC) in its special publication for the guidelines of interlocking concrete block pavement. The catalogue is designed to provide the ICBP pavement section thickness based on different traffic and subgrade strength in terms of California Bearing Ratio (CBR). The design catalogue shown in Table 5.1 provides the block thickness, bedding sand thickness, and granular subbase thickness based on the traffic conditions and subgrade CBR. The traffic considered for the design ranges from cycle track to commercial vehicle ranges from 20–50 MSA. The minimum subgrade CBR recommended for the design of ICBP is

TABLE 5.1 Design catalogue[1]

TRAFFIC AND ROAD TYPE	*TYPES OF LAYERS*	*SUBGRADE CBR (%)*		*GRADE OF BLOCK*
		ABOVE 10	*5–10*	
Cycle Tracks, Pedestrian Footpaths	Block Thickness Sand Bed Granulated Sub-base	60 mm 30±5 mm 200 mm	60 mm 30±5 mm 200 mm	M-30
Commercial Traffic Axle Load Repetitions less than 10 MSA. Residential Streets	Block Thickness Sand Bed WBM/ WMM Base Granulated Sub- Base (GSB)	60–80 mm 30±5 mm 250 mm 200 mm	60–80 mm 30±5 mm 250 mm 200 mm	M-40
Commercial Traffic Axle Load Repetitions 10- 20 MSA. Collector Streets Industrial Streets, Bus and Truck Parking Areas.	Block Thickness Sand Bed WBM/ WMM Base GSB	80–100 mm 30±5 mm 250 mm 200 mm	80–100 mm 30±5 mm 250 mm 250 mm	M-40
Commercial traffic (Container yard and seaports) Axle Load Repetitions 20-50 MSA., Arterial Streets.	Block Thickness Sand Bed WBM/ WMM Base Or WBM/WMM Base and DLC over it Granulated Sub-base	100– 120 mm 30±5 mm 250 mm 150 mm 100 mm 200 mm	100– 120 mm 30±5 mm 250 mm 150 mm 100 mm 200 mm	M-50

5%. The subgrade CBR considered for the design catalogue are 5%, 10%, and CBR greater than 10%. The minimum compressive strength recommended for IPB based on traffic conditions ranges from 30 MPa to 50 MPa. The thickness of the bedding sand is kept as 30 mm. The design catalogue aids in obtaining the design thickness of ICBP directly based on traffic and subgrade CBR.

5.2.2 Design of ICBP – Empirical Approach

The empirical design of ICBP involves evolution of design chart or design methodology based on experimental test results. There are various empirical design approaches for ICBP developed by different researchers. Rollings[2] had developed an empirical design methodology of ICBP for light and heavy traffic pavements. The ICBP design is the modification of design curves adopted by the U.S. Army Corps of Engineers. The design is based on the field testing conducted on three different test sections. The test sections were constructed to a length 6.1 m and width of 4.6 m. The main aim of conducting the test is to identify the performance of ICBP in clay when subjected to the passage of heavy military tankers. The subgrade used for the three test sections were 600mm of clay soil of type CL. CH soil is placed over the laid subgrade to maintain the CBR of 3%. The first section is laid with Z-shaped blocks with 80mm thickness placed over 100mm crushed limestone base. The second section comprises 80mm rectangular blocks laid over 100mm crushed limestone. The third section consists of 60mm unistone over 100mm gravel base. The bedding sand is laid with 25mm thickness. The test results proves that the thickness of IPB increases with the increase in the applied load. It is also concluded that the Corps of Engineers design method can become suitable for the design of ICBP. Based on the test result, the ICBP is considered as an eligible pavement for withstanding heavy load when constructed under poor subgrade.

According to the findings of Knapton,[3] the design of IPB and the bedding sand is considered as equivalent to 160mm thickness of conventional flexible pavement and base course. For the Corps of Engineers method, the equivalent thickness is considered as 165mm. The Corps of Engineers method adopt design index for the determination of the base thickness. The design index quantifies the traffic load in terms of ESAL. The thickness of the IPB is determined based on the design index. The empirical design of ICBP involves deriving the thickness of base course from the conventional design curve and subtracting the equivalent thickness of 165mm which will provide the required base and subbase for the ICBP. The base thickness is selected based on the field CBR and design index. The minimum base thickness to be adopted for ICBP is 102mm. The base thickness of ICBP is derived from the design curve based on CBR value and design Index and deducted with the equivalent thickness. The thickness of the block is derived from the design index and equivalent single axle wheel of 80 kN. The bedding sand thickness for all design index is considered to be 25mm. The suggested minimum thickness of block is 60mm for design index equivalent to one and the thickness of block is considered as 80mm for design index ranging from 2 to 8. The block thickness is increased to 100mm for design index 9 and 10. The empirical method gives an instant design of the ICBP section based on wheel load and traffic, and it is proven to be a conservative design approach.

5.2.3 Mechanistic – Empirical Method

Method I

The mechanistic-empirical design approach involves both the experimental and theoretical analysis of different mechanical characteristics of the ICBP. There are various methods of ICBP design based on mechanistic-empirical approach as the ICBP follows the flexible pavement design in many countries. Two such design methodologies are presented here. The first method of designing the ICBP is based on ASCE 58-16 guidelines.[4] The guidelines are prepared by the Transportation & Development Institute of the American Society of Civil Engineers along with Interlocking Concrete Paver Institute (ICPI). The guidelines follow American Association of State Highway and Transportation Officials (AASHTO) guide for design of flexible pavement.[5] The design chart is prepared based on AASHTO design of flexible pavement. The soils are classified in eight categories based on drainage conditions. The resilient modulus of the different categories of soil is calculated based on the CBR value. The drainage conditions are classified as good, fair, and poor based on the time to water drain. The design chart is developed in accordance with distinct traffic indexes for three different drainage conditions and eight different soil categories. The traffic index is determined as per Equation 5.1,[4] based on Equivalent Single Axle Load (ESAL). The design charts are provided individually for granular base, asphalt treated base, cement treated base and asphalt concrete base.

$$TI = 9.0 \times \left(\frac{ESAL}{10^6} \right)^{0.119} \tag{5.1}$$

Where, TI = Traffic Index
ESAL = Equivalent Standard Axle Load
The step adopted for the design of ICBP is as follows:

Step I: Determine the ESAL and traffic index as per the equation
Step II: Determine the drainage condition
Step III: Conduct sieve analysis and conclude the subgrade category
Step IV: Select thé type of base
Step V: From the design chart determine the ICBP pavement section based on ESAL, traffic index, drainage condition, subgrade category and type of base material.

Method II

The mechanistic-empirical design approach of ICBP is developed based on rutting and deflection failure criteria.[6,7] The shear and deflection studies are carried out using large scale direct shear test and plate load test. All the geometric properties of the ICBP such as block shape, block thickness, block compressive strength, block laying pattern, block laying angle, loading position, block joint width, gradation of bedding, and jointing sand are considered for the deflection and stress analysis. The influence of supporting layers such as bedding sand thickness, and granular layer thickness are also analysed. Statistical analysis are carried out to determine the most influential factors among the considered parameters. Two deflection models were developed for zigzag shape with stretcher and herringbone pattern. Rutting above the subgrade due to vertical compressive strain and deflection of the ICBP surface are considered as the failure criteria. Design charts are prepared for different traffic conditions and subgrade CBR as shown in Table 5.3 based on the design procedure explained below. The step by step design procedure is as follows:

Step I: Input Parameters

The input value consists of a number of traffic load repetitions in terms of Million Standard Axle (MSA) and was derived from Equation 5.2. The input parameters were granular/base thickness (H_2), block thickness (h_1), bedding sand thickness (h_2), wearing layer thickness (H_1). The wearing layer thickness was calculated as the sum of block thickness and bedding sand thickness. The other input parameters considered were pressure (P_0), radius of contact area (R_0), CBR (%), type of pattern, plan area and perimeter of the block, and actual vertical strain (ε_v). The actual strain was determined as per the procedure of IRC 37.[8] The actual strain was calculated based on the IITPAVE software which is programmed by IIT Kharagpur based on the conventional method of finding three-layer stress, strain and deflection. The minimum growth rate adopted for the design was 5%. The lateral distribution factor and vehicle damage factor were calculated as per the procedure specified in IRC 37.[8] The design traffic is determined from Equation 5.2.

$$\mathbf{N = A \times D \times F \times \frac{365 \times \left[\left((1+r)^{n-1}\right)\right]}{r}} \quad (5.2)$$

Where, N = Cumulative number of standard axles during the design period, A = Commercial Vehicle per day, D = lateral distribution factor, F = vehicle damage factor (VDF), n = design period in years, r = annual growth rate of commercial vehicles.

Step II: Determining Material Properties

The basic properties considered for the design of ICBP were elastic modulus of the wearing layer, modulus of base/granular layer, and the modulus of subgrade. The flexible pavement design prefers resilient modulus of the base and subgrade, which is a ratio of deviator stress by recoverable strain obtained from the triaxial test. The testing of resilient modulus is expensive as the conventional triaxial test may not support arriving at the properties. In order to overcome this difficulty, IRC 37[8] has recommended empirical formulae in deriving the resilient/elastic modulus of the subbase and subgrade, which is given in Equations 5.3 to 5.6. The elastic modulus of the wearing layer is provided in Equation 5.8. The wearing layer of ICBP consists of IPB, jointing sand, and bedding sand. The shape factor is the geometric properties of IPB. The shape factor is the ratio of vertical surface area to plan area as shown in Equation 5.7. Vertical surface area is a product of the perimeter of the block by the thickness of the block.

$$\mathbf{E_{SG} = 10 \times CBR \text{ if the } CBR \leq 5\%} \quad (5.3)$$

$$\mathbf{E_{SG} = 17.6 \times (CBR)^{0.64} \text{ if the } CBR > 5\%} \quad (5.4)$$

$$\mathbf{E_{Gran} = 0.2(h)^{0.45}{}_{x\,ESG}} \quad (5.5)$$

$$\mathbf{E_{wear} = 418.688 + 90.85\,H + 2.528\,E_{Gran}} \quad (5.6)$$

$$\mathbf{S.F = \frac{Vertical\,Surface\,Area\,(m^2)}{Plan\,Area\,(m^2)}} \quad (5.7)$$

Where, E_{SG} – Elastic modulus of subgrade (MPa), CBR -California Bearing Ratio (%), E_{Gran} – Elastic modulus of the granular layer, h – Thickness of granular layer (mm), E_{wear} – Elastic modulus of wearing layer, S.F – Shape Factor

Step III: Calculating Allowable Deflection

The allowable deflection is the limited deflection for the assumed pavement section. Sun[9] proposed a rut depth model developed with an accelerated pavement tester. The rut depth limit for the ICBP suggested was 30mm because of the presence of bedding sand. As per IRC 37, the allowable rut depth is 20mm. Hence for the present study, 25mm was considered as the limited rut depth for ICBP pavement section. The proposed model was designed for a maximum load of 400 kN considering the loads encountered in the port area. For the present study, a maximum of 50kN is proposed based on the maximum standard axle load of 80 kN with a configuration of a single axle dual wheel on both sides. Therefore, based on the pilot study, a reduction factor of 0.80was applied to the rut depth equation. The rut depth model in terms of deflection is shown in Equation 5.8. The motive was to calculate the limited or allowable deflection, which causes a rut depth of 25mm.

$$0.24D_0 - \ln(0.7D_0) = \ln(961) - \ln(R.D) - 0.16 \times \ln(P_0) - 2.21x \ln(R_0) - 0.11 \times H_1 + 0.265 \times \ln(N) \quad (5.8)$$

Where, D_0 – Allowable deflection (mm), R.D – Rut Depth (mm), P_0 – Pressure (MPa), R_0 – radius of contact area (cm), H_1 – Thickness of wearing layer (cm), N – Cumulative standard axle (MSA)

Step IV: Calculating Actual Deflection

The actual deflection was calculated based on the developed model as shown in Equations 5.9 and 5.10. The actual deflection is a function of shape factor, applied pressure (P_0), elastic modulus of the granular layer (E_{Gran}), elastic modulus of subgrade (E_{SG}). The elastic modulus of granular layer is a function of the thickness of base layer. The elastic modulus of subgrade is a function of the California Bearing Ratio (CBR). The actual deflection should be lower than the allowable or limited deflection. If the deflection check is not satisfied, then the section has to be revised. The initial revision was done by increasing the granular layer to a maximum of 550mm. Then the actual and allowable deflection was checked, and if not satisfied, then the thickness of the block should be increased. The trial and error should be continued till the calculated actual deflection was less than the allowable deflection.

$$\text{Deflection} = (0.787 - 0.089SF - 0.001E_{Gran} - 0.004\,E_{SG} + 1.086P) \times C \quad (5.9)$$

where, SF – Shape factor, E_{Gran} – Elastic modulus of base layer E_{SG} – Elastic modulus of CBR. C- correction factor, C=1 for load up to 40kN and C = 0.98 for load greater than 40kN.

$$\text{Deflection} = (0.842 - 0.089SF - 0.002E_{Gran} - 0.004E_{SG} + 1.335P) \times C \quad (5.10)$$

where, $E_{Gran} = 0.2*h^{0.45}*E_{SG}$, $E_{SG} = 10 * CBR$ for CBR≤ 5, E_{SG} = 17.6 * (CBR)0.64for CBR > 5 %, h = Thickness of granular layer (mm), CBR = California Bearing Ratio (%)

Step V: Calculating Actual and Allowable Vertical Strain

The vertical compressive strain is considered a critical factor for the rutting of subgrade. The actual vertical strain value is an input parameter that is determined from IITPAVE software exclusively derived for the determination of stress, strain, and deflection based on three-layer theory. The input parameters of the IITPAVE are pressure, radius of contact area, elastic modulus of wearing layer, granular layer, and subgrade, thickness of wearing layer, and base or granular layer. The vertical compressive strain at the top of the subgrade is considered as the actual vertical strain. Equations 5.11 and 5.12 were suggested for the determination of vertical compressive strain as specified in IRC 37. The 80% reliability was considered when the design traffic is less than 20 MSA, and 90% reliability was considered when the design traffic was greater than 20 MSA. The allowable vertical compressive strain was a function of the number of traffic load repetitions. The allowable strain computed was checked with the actual vertical strain value. When the allowable strain was lower than the actual strain, the design consideration is safe; if not, the section needs to be revised.

$$N = 4.1656 \times 10\text{-}08\, [1/\varepsilon_v]^{4.5337} \text{ for 80 \% reliability} \quad (5.11)$$

$$N = 1.4100 \times 10\text{-}08\, [1/\varepsilon_v]^{4.5337} \text{ for 90 \% reliability} \quad (5.12)$$

Where, N – Number of cumulative standard axle (MSA), ε_v – Vertical compressive strain

Step VI: Drainage Properties

The drainage properties of the ICBP pavement are vital as the wearing layer is pervious, which allows the drainage of the stormwater into the pavement section. The provision of a better drainage layer below the bedding sand will improve the drainage properties of ICBP. Therefore, the drainage characteristics were classified based on the CBR of the subgrade as per the recommendations and specifications of IS 1498.[10] Based on the drainage classification, provision on the thickness of the drainage layer is recommended in the granular layer. The classification of the drainage properties based on CBR is shown in Table 5.2. The selection of the subbase in the granular layer was based on the drainage properties. The selection requires a minimum base requirement of 150mm, and the subbase layer was suggested to be 150mm, 200mm, and 250mm based on the drainage condition. Gradation III and IV, as per the MoRTH specifications

TABLE 5.2 Drainage properties[7]

DRAINAGE CONDITIONS	*SOIL CLASSIFICATION*	*CBR (%)*	*DRAINAGE SUB-BASE LAYER (MM)*
Excellent	GW, GP, SW, SP	>15	150
Fair	GM, GC, SM, MI, ML	10–15	200
Poor	SC, CL, OI, MH, CH	5–10	250

Notes: G = gravel, W = well graded, P = poorly graded, M = silt, C = clay, S = sand, L = low compressibility, I = intermediate compressibility, H = high compressibility

were considered as the subbase layer if the thickness is greater than or equal to 200mm.[11] For the subbase layer of 150mm thickness, a single drainage cum separation layer of gradation V and VI as per MoRTH specification was preferred. The base layer can be Wet Mix Macadam (WMM) or Water Bound Macadam (WBM). The design chart prepared based on the design procedure is shown in Table 5.3.

5.3 DESIGN OF ICBP – PORTS

The ICBP is found to be a constructive material in the field of pavement engineering. The unique advantage such as easy laying, better strength, and easy maintenance promotes the use of ICBP for pavements in ports which deals with heavy loads. The research on the development of ICBP on ports was initiated during the 1970s by Knapton. He also developed the British Ports Association Heavy Duty Pavement Design Manual,[12] which illustrates the design methodology of ICBP in port areas. The ICBP is laid in Europe Container Terminus, Rotterdam, Netherland which handles some of the largest containers in the world. The ICBP is laid over an area of one million square meters on the Rotterdam Port. The developed chart aids the determination of the pavement thickness for a wheel load of 30 tons. Since the IPB is available in standard thickness with unique interlocking mechanism, the design focused on the thickness of the base course with different materials. The IPB holds a higher stiffness value range from 1000 MPa to 5000 MPa. The stiffness of IPB exhibits a lower impact of less than 2% in the stress values at base course. The design for the port assumes that the geometric properties of IPB plays a negligible role on the applied load.

The rectangular IPB with a plan dimension of 200mm × 100mm is adopted in order to provide a stable surface. The thickness of the IPB for the design is taken as 80mm with bedding sand thickness as 30mm. The total thickness

TABLE 5.3 Design chart[7]

ZIGZAG – HERRINGBONE

CBR (%)	ICBP PAVEMENT SECTION	CUMULATIVE STANDARD AXLE (MSA)						
		0.5	1	2	5	10	20	50
5	Paver + Bedding	90	90	110	110	130	150	150
	Unbound Base layer	225	275	250	300	150	150	350
	Unbound Sub-base				250	200	200	250
7	Paver + Bedding	90	90	110	110	130	150	150
	Unbound Base layer	175	225	200	250	150	150	300
	Unbound Sub-base				200	150	150	200
9	Paver + Bedding	90	90	110	110	110	130	150
	Unbound Base layer	150	175	200	250	300	350	150
	Unbound Sub-base				200	200	200	150
11	Paver + Bedding	90	90	110	110	110	130	150
	Unbound Base layer	150	150	200	250	150	250	150
	Unbound Sub-base					150	150	150
13	Paver + Bedding	90	90	110	110	110	130	130
	Unbound Base layer	150	150	200	200	250	250	300
	Unbound Sub-base							150
15	Paver + Bedding	90	90	110	110	110	130	130
	Unbound Base layer	150	150	200	200	200	200	200
	Unbound Sub-base							150

ZIGZAG – STRETCHER								
		CUMULATIVE STANDARD AXLE (MSA)						
CBR (%)	ICBP PAVEMENT SECTION	0.5	1	2	5	10	20	50
5	Paver + Bedding	90	90	110	110	130	150	150
	Unbound Base layer	225	275	250	350	250	150	400
	Unbound Sub-base				250	200	200	250
7	Paver + Bedding	90	90	110	110	130	150	150
	Unbound Base layer	175	225	200	250	150	150	300
	Unbound Sub-base				200	150	150	200
9	Paver + Bedding	90	90	110	110	110	130	150
	Unbound Base layer	150	175	200	250	300	350	150
	Unbound Sub-base				200	200	200	150
11	Paver + Bedding	90	90	110	110	110	130	150
	Unbound Base layer	150	150	200	250	150	250	150
	Unbound Sub-base					150	150	150
13	Paver + Bedding	90	90	110	110	110	130	130
	Unbound Base layer	150	150	200	200	250	250	300
	Unbound Sub-base							150
15	Paver + Bedding	90	90	110	110	110	130	130
	Unbound Base layer	150	150	200	200	200	200	200
	Unbound Sub-base							150

for the wearing layer of ICBP is considered as 110mm. The design chart for the ICBP for the port areas is developed based on numerical simulations on the different pavement components. The analysis is carried out by considering as linear elastic model. The design chart is developed by considering 110mm of wearing layer and Cement Bound Granular Mixtures (CBGM) base layer. The properties considered for the analysis and simulation different layers are elastic modulus and Poisson's ratio. The elastic modulus for the development of design chart considered for pavers, base, subbase and capping are 4000 MPa, 40000 MPa, 500 MPa and 250 MPa respectively. The elastic modulus of the subgrade is based on the subgrade CBR.

The thickness of the base layer is determined from the developed design chart based on the single equivalent wheel load and the repetitions of load. The determined base thickness is the thickness of CBGM. When other alternate base materials are used, the manual suggests Material Equivalence Factor (MEF) to convert the equivalent base thickness. The MEF for the CBGM is one. For instance, if C16/20 graded base course is used, then the thickness of the base layer determined from the design chart is multiplied by 0.79. The minimum base thickness recommended for ICBP is 200mm. The design manual aids in arriving the base thickness for different alternative base materials, the thickness of the IPB is kept as 80mm and bedding sand thickness as 30mm.

5.4 DESIGN OF ICBP – AIRPORTS

The ICBP are used in airports as taxiway, hanger, and apron pavement due to its high structural performance. The unique application of ICBP is well witnessed in airport pavements which is illustrated in detail in Chapter 7. There are different methods adopted by the researchers for the design of airport ICBP. The modified Federal Aviation Administration (FAA) design is discussed for ICBP design on airport pavements.[13] The airport ICBP design thickness is mainly influenced by pavement material characterization, subgrade strength, intensity of wheel load and traffic. The additional care to be considered for airport ICBP design includes resistance to fuel spillage, stability and skid resistance. In this regard, concrete pavers have demonstrated proven performance for a wide variety of airport applications as the flexible pavement is more prone to damage on fuel spillage. The rigid pavements are used in most of the hanger, apron areas. However, the rigid pavement is subjected to joint sealing failures. These two major disadvantages from the flexible and rigid pavement promotes the application of ICBP in airport pavements other than runway as the blocks are subjected to disintegrated due to the high propulsion of jet engines.

The design procedure of airport ICBP is similar to the FAA pavement design approach. The total thickness of the pavement is arrived at based on the FAA method. The FAA method assumed that the airport pavements are designed with maximum of 200 to 250mm of bituminous concrete. The FAA design of ICBP is determined by replacing the thickness of the asphaltic concrete layer with the thickness of Interlocking Paver Block (IPB) and bedding sand. The ideology behind the replacement of entire asphalt concrete with the thickness of IPB and bedding sand is that the IPB and bedding sand will be strong and capable of performing similar to that of the asphalt concrete. The design chart from FAA Advisory Circular 150/5320-6D[14] is used to determine the total thickness of the pavement based on aircraft weight and CBR. This advisory circular is cancelled by FAA and alternately, FAA Rigid and Flexible Iterative Elastic Layer Design (FAARFIELD) software developed by FAA aids in determining the total pavement thickness of the pavement based on CBR and aircraft weight as per the revised FAA Advisory Circular 150/5320-6E.[15] The determined pavement thickness of the asphaltic pavement is converted to the thickness of IPB and bedding sand. The minimum base thickness for the airfield pavement is provided as 300mm. The design of ICBP airports follows the FAA design by replacing the asphaltic layer by the IPB and bedding sand thickness.

5.5 DESIGN SOFTWARE

The design of ICBP is made easier with the use of design software. There is various research carried out to evolve the design software exclusively for the ICBP. Few such software are LOCKPAVE PRO™, BLP3D, NITT-BLOCKPAVE etc. These software packages are exclusively developed for the structural design of ICBP. LOCKPAVE PRO is developed by Favor Langsdorff Licensing Limited and published in US and Canada. LOCKPAVE PRO is developed for the designers and engineers to provide the ICBP section based on subgrade soil, traffic, pavement material characterisation, and environmental factors. The BLP3D is unique tool in determining the shear modulus of joints and spring coefficients of bedding sand. However, the BLP3D is not commercialised due to the deviation in the fundamental concept of designing the structural model and calculations. The

DesignPave is another software program developed for the design of ICBP by Concrete Manufacturing Association of Australia (CMAA) in collaboration with the University of South Australia.[16] The thickness of the ICBP is computed based on material properties and applied load using the Method of Equivalent Thickness (MET). The final report gets generated automatically with the design analysis and cost analysis. Yet another software for the design of ICBP is NITT-BLOCKPAVE.[6] C# scripts were used to develop the NITT-BLOCKPAVE software. C# Windows Forms were used to create the user interfaces, while C# Mathematical Functions were used to perform the calculations and reports. The major priority of the software is to reduce the computation time and to ensure that the design section satisfied the deflection and rutting requirements. The software calculates the actual and permissible deflection. If it is satisfied, the design was considered safe, else; the iteration will continue until the actual deflection is less than the permissible deflection. The rutting criteria were also tested by comparing the actual vertical compressive strain to the permissible vertical strain, and iteration would continue until the check is fulfilled. In addition to the software developed exclusively for ICBP, few other common Finite Element Analysis (FEA) software such as ANSYS, ABAQUS, SAP and PLAXIS are used for the determination of stress, strain, and deflection properties of ICBP. The design software developed exclusively for ICBP design enables the user to compute the thickness of ICBP rapidly.

5.6 SUMMARY

The application of ICBP is widespread in its use on light traffic and heavy traffic highway pavements. The ICBP extends its application to ports and airports. Compared to highway pavements, the load on the port and airport pavements are heavy. Various research is carried out to determine the ICBP design for pavements. The design of ICBP is based on catalogue, empirical approach, and mechanistic-empirical approach. The design of ICBP mainly depends on the failure criteria based on shear, rutting and deflection. The design of ports and airports follows the conventional method of flexible pavement design. The thickness of the ICBP wearing layer is arrived at based on the equivalency factor for the determined bituminous layer thickness of the flexible pavement. The design software is also available exclusively for determining the ICBP section. The mechanical properties of ICBP can also be evaluated based on the common FEA software.

REFERENCES

1. IRC SP 63 (2018) Guidelines for the Use of Interlocking Concrete Block Pavement. *Indian Road Congress Special Publication*, New Delhi.
2. Rollings (1984) Corps of Engineers Design Method for Concrete Block Pavements. *Proceeding of the 2nd International Conference on Concrete Block Paving*, Delft, pp. 147–51.
3. Knapton, J. (1976) The Design of Concrete Block Roads, Report No. 42.515. *Cement and Concrete Association, Wexham Springs*, United Kingdom.
4. ASCE 58-16 (2016) Structural Design of Interlocking Concrete Pavement for Municipal Streets and Roadways. *American Society of Civil Engineers*, Virginia, United States.
5. AASHTO (1993) Guide for Design of Pavement Structures. *American Association of State Highway and Transportation Officials,* United States.
6. Arjun Siva Rathan, R.T., Sunitha V., and V. Anusudha (2021) Development of Design Procedure for Interlocking Concrete Block Pavement. *International Journal of Pavement Engineering*, 23(14), 5015–5029. https://doi.org/10.1080/10298436.2021.1990290
7. Arjun Siva Rathan, R.T., and V. Sunitha (2021) Development of Deflection Prediction Model for the Interlocking Concrete Block Pavements. *Transportation Research Record: Journal of the Transportation Research Board,* 2676(3), 292–314. https://doi.org/10.1177/03611981211051339
8. IRC 37 (2018) Guidelines for the Design of Flexible Pavements. *Indian Road Congress*, New Delhi.
9. Sun, A. L. (1996) Design Theory and Method of Interlocking Concrete Block Paving for Port Areas. *Fifth International Conference on Concrete Block Pavement*, Tel Aviv, Israel.
10. IS 1498 (1970). Classification and Identification of Soils for General Engineering Purposes. *Bureau of Indian Standards*, New Delhi.
11. MoRTH (2013) Specifications for Roads and Bridge Works, *Ministry of Road Transport and Highways,* New Delhi.
12. John Knapton (2009) British Ports Association Port and Heavy Duty Pavement Design Manual, *9th International Conference on Concrete Block Paving*, Buenos Aires, Argentina.
13. McQueen, Roy D., Knapton, J., Emery, J., and David R. Smith (2012) Airport Pavement Design with Concrete Pavers – A Comprehensive Guide. *Interlocking Concrete Paver Institute*, United States.

14. FAA Advisory Circular 150/5320-6D (1995) Airport Pavement Design and Evaluation. *Federal Aviation Administration,* Washington DC, United States.
15. FAA Advisory Circular 150/5320-6E (2009) Airport Pavement Design and Evaluation. *Federal Aviation Administration,* Washington DC, United States.
16. CMAA (2014) Pavements – Design Guide for Residential Accessways and Roads. *Concrete Manufacturing Association of Australia,* Australia.

Construction Process and Quality Control of ICBP

6

6.1 INTRODUCTION

The pavement materials need to be properly inspected to ensure that they meet the necessary qualities as per the standard specifications. To ensure that the ICBP will function more effectively, its construction process must be carried out in an accurate and precise manner. The ICBP construction process involves subgrade preparation, base and subbase construction, edge restraint installation, bedding sand preparation, paver block laying, and jointing sand filling. Manual labour and machinery are both utilised in the process of laying down paving blocks. The wearing layer is compacted using a plate load equipment and a rubber tyred roller. The degree of compaction is ensured by the use of sand replacement and core cutter methods. In regions where the alignment altered, a different construction procedure was used. Large Scale direct shear test, Benkelman Beam tests, and other methods can be used to determine the quality of the laid ICBP. The tests need to be carried out in routine to evaluate the performance of ICBP.

6.2 COMPONENTS OF ICBP

The ICBP consists of wearing layer, base layer, subbase layer and subgrade. The wearing layer comprises of Interlocking Paver Block (IPB), bedding sand

DOI: 10.1201/9781003432371-6

and jointing sand. The base layer and subbase layer materials are similar to that of the flexible pavements. The components of the ICBP are as follows:

- Interlocking Concrete Block
- Bedding Sand
- Jointing Sand
- Base Layer
- Subbase Layer
- Subgrade

The materials specification and properties of these materials are explained in Chapters 2 and 3.

6.3 CONSTRUCTION PROCESS

The construction process for the Interlocking Concrete Block Pavement (ICBP) is carried out in the following stages:[1]

A. Subgrade compaction
B. Construction of sub-base and base layers
C. Fixing of edge restraints
D. Construction of bedding layer
E. Laying paving blocks and compaction
F. Filling the jointing sand and compaction

6.3.1 Subgrade Preparation

The supporting layers and their functions of the ICBP are similar to that of the flexible pavement. The subgrade is considered as the lower most supporting layer and is found to be the significant component in the performance of ICBP. The failures in the subgrade shall lower the design life of ICBP. The minimum length of 100 m is considered for the preparation of the subgrade and can be proceeded to the entire length. The subgrade can be laid 300mm wider from the layout to achieve full compaction. The depth of water table should be 600mnm below the subgrade layer. The subgrade should be free from loose soil and if encountered with loose soil, the subgrade should be scarified to a minimum depth of 150mm. The minimum thickness of the subgrade should be compacted to 500mm. The compaction of each layer shall be of 150mm.[2] The subgrade is compacted within the ±2% Optimum Moisture Content (OMC) determined from the laboratory standard or modified proctor test.[3] The moisture content in the field can be evaluated immediately using rapid moisture meter. The dry density of the compacted subgrade should not be less than the 98% laboratory Maximum Dry Density (MDD) determined from standard or modified proctor test. The dry density in field can be measured using sand cone replacement method or nuclear gauge. After compaction, the surface evenness should not be more than 15mm when measured with a 3m straight edge. The subgrade should have a CBR not less than 5% as a minimum criterion for the design of ICBP.[4]

6.3.2 Construction of Base and Subbase

Base and sub-base courses are laid over the subgrade to serve as a separation and drainage layer. The different types of base and subbase layer are Wet Mix Macadam (WMM), Water Bound Macadam (WBM), Crusher Run Macadam, cement treated base, lime treated and bituminous treated base. The minimum layer thickness for the granular base layer is 100mm. The thickness of bound base layer is 67% the thickness of the granular base layer subjected to a minimum thickness of 100mm. If the subbase is of grading IV and V as per MoRTH Specification, a single layer of 200mm shall be provided.[2] The compaction of the laid base layer is carried out by rollers with ±2% Optimum Moisture Content (OMC) determined from the laboratory standard or modified proctor test. The density of the compacted layer is tested and required to meet the 98% laboratory dry density. Rolled lean concrete shall be provided for cement based base layers.[5] The methodology of providing the materials differ based on the different types of base layer. The WBM layer possesses a procedure of laying stones and placing grade soil over the laid stone and compacted, while WMM is a mixture of aggregate and filler on required gradation, prepared from the mix plant. However, the surface level of the base layer needs to be taken care of without any irregularity. The level difference of base layer significantly affects the uniformity of the Interlocking Paver Blocks.

6.3.3 Fixing of Edge Restraints

The ICBP wearing layer is discrete and is expected to have a larger deformation during the braking and maneuvering of vehicles. The edge restrained stones need to be constructed to counteract the deformation and to enhance the interlocking efficiency. The edge restraint can be laid during the base layer construction. The edge restraint is manufactured using precast blocks or casted as in situ. The edge restraint should have a minimum strength to resist the confining pressure. The edge restraint should possess compressive strength of 30 MPa and flexural strength of 3.8 MPa.[1] The vertical face of the edge restraint should be placed inside blocks. The compaction of the blocks using rollers at the edge region of the restraint blocks is challenging and therefore manual compactor needs to be used in that region. The IPB when laid near to the edge restraint may require cut block pieces. This may result in increase in joint gaps or non-uniform gaps between the block and edge restraint which may reduce the interlocking efficiency. It is required to seal the gaps between the edge restraint and the corner blocks with cement mortar with a ratio of 1:3.

6.3.4 Placing and Screeding of Bedding Sand

The bedding sand should be prepared for the desired gradation. The compaction of the bedding is one of the the significant criteria to be taken care of in the construction of bedding sand. The specified required thickness for the bedding sand ranges from 20 to 30mm so as to reduce the risk of depression due to the applied load which results in surface unevenness in ICBP. The OMC used for the bedding sand ranges from 6 to 8%. The bedding sand is placed to a loose thickness of 25–50mm. The compaction preferred for the ICBP is by an earth vibratory compactor with two to three passes. The earth vibratory should have a minimum weight of 0.6 tons or more.[1] After the completion of the vibratory compaction, the bedding sand is levelled with screed boards to the required thickness. It is not advised to step in once the bedding sand is levelled and compacted.

6.3.5 Laying of Blocks

The IPB shall be laid above the bedding sand. The IPB is laid either by manual method or mechanized mode.

6.3.5.1 Manual Laying

The manual laying requires skilled labour capable of maintaining the bedding sand level and to handle the blocks. The work is initiated and moves backwards so that the labourer can have a full view of the completed work and it also aids to maintain the uniformity of laying. It is estimated that one pavior can complete 50 to 75 m^2 of paving per day.[1] The size of the block plays a significant role on the efficiency of laying. The size of the block should be chosen with a mean length to mean width ratio between 1 and 3. The maximum horizontal dimension of the block shall be restricted to 280mm and maximum thickness shall be restricted to 140mm. The size and weight of the block should be easy for a pavior to handle in one hand. The joint width needs to be maintained uniformly while laying the IPB. The non-uniform joint width may result in poor performance of ICBP. The joint width can be measured at a series of randomly selected intervals during laying to check the consistency in the joint width. The measurement can be determined using a calibrated steel mandrel or Vernier caliper at random locations and analysed using a statistical method.

6.3.5.2 Machine Laying

The machine laying uses special equipment manufactured exclusively for the laying of blocks. The machine laying process is semi-automatic and automatic. The hand operated equipment is capable of laying paving block cluster of area ranges from 0.3 to 0.5 m^2. The fully mechanized equipment can handle paving block cluster up to 1.2m^2. The other mode of laying paver blocks is through an automatic interlocking paver machine. This is an advanced machine which saves time when laying and the human effort required. The machine is capable of laying paver blocks of different shapes and laying patterns. The paving machine moves along the straight line and the blocks are kept on the chamber. The machine is capable of aligning the blocks in a definite laying pattern and paved with precise quality maintaining the uniform joint width. The machine is designed to pave the blocks of about 500 to 700 sqm per day. The width of the paver machine can be adjusted and can accommodate the width of the pavement. The automatic paver machine is inefficient while laying the paver blocks in a curve region. In such cases a combined manual laying and automatic paver laying is suggested with quality control.

6.3.6 Filling the Jointing Sand and Compaction

The jointing sand is an important component in ICBP which aids in the shear transfer of ICBP. The jointing sand of the required gradation is placed over the IPB at a few intervals. The bedding sand is then broomed to the jointing sand. The compaction is needed after placing the jointing sand. The compaction is carried out using standard vibratory compactors with a weight of 90kg over a plate area of 0.3 m^2 with a centrifugal force of about 15 kN. The compaction for heavy traffic is 300–600 kg over a plate area of about 0.5–0.6m^2 with a centrifugal force of about 30–65 kN[1]. The compaction can be carried out with two to six passes of a vibratory roller with rubber coated drums to avoid damage to the IPB. The effective compaction aids in the better placement of IPB without undulations and proper settlement of jointing sand. After the compaction, the surface is completely cleaned. The traffic can be opened immediately after the compaction of the jointing sand and cleaning process. Fourteen days' curing for lime or cement treated base layers is required.

6.3.7 Laying Procedure

The Interlocking Paver Blocks (IPB) have different shapes, sizes, and laying patterns. The following are some of the points to be considered during the laying of (IPB):[1]

- The IPB when laid in slope region should be initiated from downhill and move towards the uphill in order to avoid the slipping of blocks.
- For an irregular outline the string should be placed at the straight centre line and start moving towards the edge restraints as shown in Figure 6.1.
- Broken blocks are not permitted to be used other than for connections and edges of the block subject to the blocks being broken using a block splitter.
- The laying of blocks should be precise to maintain camber for the drainage of storm water.
- The laying pattern needs to be realigned at curves to avoid weaker joints. A herringbone pattern is preferred at curves and can continue on the same or other respective patterns on straight approaches.
- In areas where intrusions like drainages or manholes are present, the laying should initiate simultaneously from the starting point rather than laying around the intrusion to avoid accumulation of closing errors as shown in Figure 6.2.

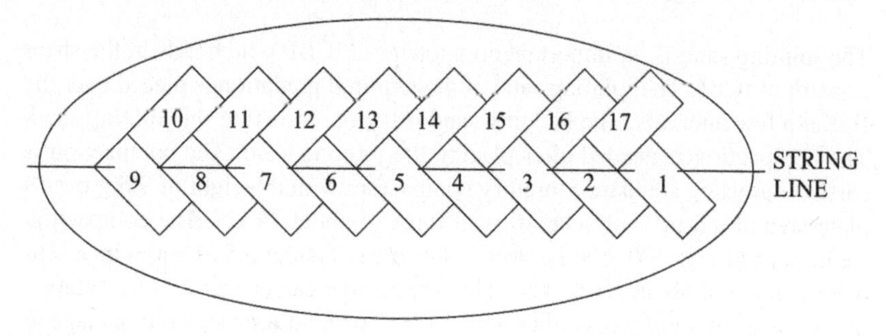

FIGURE 6.1 Block laying procedure.

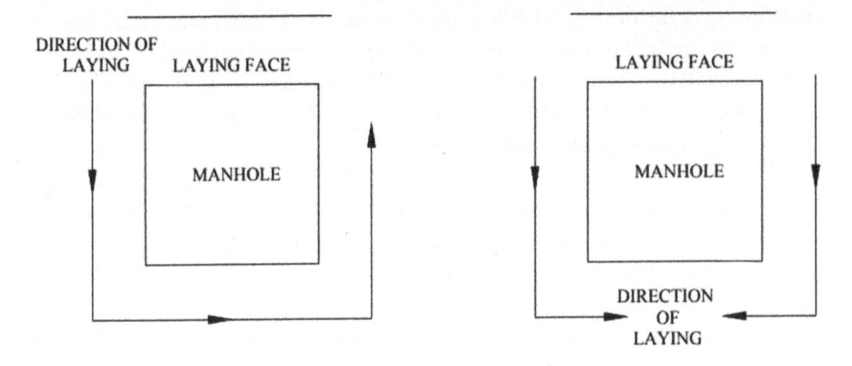

FIGURE 6.2 Block laying procedure around manhole.

6.4 QUALITY CONTROL OF ICBP – STRUCTURAL EVALUATION

Elastic deflection and rutting are the two main components of ICBP failure criteria.[6] Plate load test and accelerated pavement testing equipment are the commonly used laboratory testing equipment for the determination of the elastic deflection, and rutting. The plate load test can be carried out both in the laboratory and field. The test can be static or dynamic. The accelerated testing is done by constructing trial pit sections where the actual traffic conditions are simulated with repeated loads. Accelerated pavement testing is costlier compared to plate load testing. Most of the studies on ICBP is carried out using laboratory plate load testing.

6.4.1 Plate Load Test

The plate load test is carried out to determine the deflection profile, strain and stress of the ICBP pavement section. The plate load test comprises of a steel tank, hydraulic jack, loading frame, loading plate, and accessories to determine the deflection. The field plate load test for the ICBP deflection investigation is carried out with a pit dimension ranging from 775m × 775m to 2m × 2m.[7,8,9] The boundary effect is minimal for the ICBP with a tank size greater than 775m × 775m. The pressure cells and LVDT are the preferred instruments for measuring applied pressure and deflection during plate load test. The plate load test is conducted by applying a 5 kN seating load and gradually increasing the load, noting deflections at 5 kN loading increments. A maximum initial load of 3.5 kN was applied to base courses, and a maximum initial load of 5 kN was applied to ICBP.[10] The dynamic load on ICBP is carried out by the pulsing load for 1 million cycles at 0.7MPa pressure over a 300mm diameter steel plate.[11]

The elastic and plastic deformations were measured during the dynamic loading phase and static loads were measured using a data logger. The magnitude of the permanent deformation is greatly influenced by the block shape, jointing material, bedding material, and laying pattern. The plate load test is conducted for subgrade and CBP with and without Granular Sub-Base (GSB) and the deflection was measured by providing a hydraulic load on a 75cm diameter plate. The results show that the deflection is reduced when GSB is provided over subgrade.[7] A rubber sheet was used to simulate the subgrade layer.[12] Load cells and deflection gauges were used to determine the applied pressure and deflection of the ICBP. The plate load test used commonly for the analysis of ICBP performance employed static load. In some cases, dynamic loads were also used in the plate load experiment. The dynamic load is simulated using Light Weight Deflectometer (LWD) to perform the dynamic plate load test.[13] The primary concern in using LWD is that the maximum capacity of the LWD is only 9000N. The geometric properties and supporting layers of ICBP plays a significant factor in the determination of deflection profile and stress measurement in ICBP.

6.4.2 Accelerated Pavement Testing

The accelerated pavement testing is a more realistic experiment that simulates actual traffic behaviour with repetition of wheel movement. Furthermore, the experiment includes a rubber wheel tyre, which is more realistic than the rigid plate used in the plate load test. Accelerated loading is carried out with roughly 1.5 equivalent standard axles of 81 kN applied to the test lengths every 60 seconds.[14] Ryntathiang[15] performed an accelerated pavement test with the equipment assembly consisting of four guide wheels arranged in such a way that the entire load of the structure is shared by the dual wheel and the two rear guide wheels resting on two parallel tracks 2.33m apart. The overall weight on the wheels was set at 40kN, which equates to half of a truck's typical axle load. Woodman and Halliday[16] tested the applicability of ICBP in airfield pavements using accelerated traffic tests through an experimental pavement built and subjected to 10,000 load repetitions of a ten-tonne wheel load. It was discovered that the concrete block surfacing had little impact on the structural strength of the pavement. Although the experiment employing accelerated pavement testing gives a more realistic approach to determining rutting and elastic deflection, the equipment is heavy, expensive and requires a large working space. Because plate load tests are less expensive, they were used in most investigations to understand the behaviour of ICBP better.

6.4.3 Light Weight Deflectometer and Benkelman Beam Deflectometer

The structural evaluation with field tests is critical to understand the performance of the ICBP. The field test is conducted in two modes. The first is an evaluation of field testing on ICBP sections built and open to public traffic. The second mode involves the construction of trial road sections in open areas, where real traffic is replicated and evaluated. Benkelman Beam Deflectometer (BBD) and LWD are the field-testing tools used to evaluate the performance of ICBP. The LWD is used to determine the elastic modulus and deflection of ICBP. The LWD are intended for granular base layers and subgrade. However, the ICBP wearing layer is not continuous and rigid as compared to flexible pavement and rigid pavement, LWD is used for the evaluation of ICBP.[13] The LWD procedure is illustrated in ASTM E2583.[17] The LWD is a light, portable device that can produce impact load between 13.2 kN to 15 kN from a falling weight range from 10kg to 50 kg.[18] The height of fall of the mass ranges from 100mm to 850mm.[19] The measurement of the deflection of ICBP is carried out by dropping the weight for ten blows and the average of the last three readings are considered as the surface deflection.[20]

BBD is another remarkable equipment for the measurement of ICBP surface deflection. The method of measurement of BBD is similar to that of the flexible pavement. The test procedure of BBD follows IRC 81.[21] The deflection is measured using a dial gauge fixed in the BBD beam. The monitoring end is kept in between the two rear axle wheel of a vehicle which weighs 81 kN load. The test is carried out in closely spaced intervals of 25 to 50 m in alternate positions. The rebound deflection is calculated based on temperature and moisture corrections. The BBD and LWD are the effective methods adopted to determine the surface deflection of ICBP. A 3m straight edge was used to determine the irregularities in the longitudinal profile.

6.4.4 Large Scale Direct Shear Test

Large scale direct shear test apparatus is one of the primitive tests adopted to investigate the interface frictional behaviour of soil-structures. The test procedure was carried out as per ASTM D5321/D5321M.[22] The test set-up consists of two square boxes, each of size 300 mm × 300 mm × 100 mm.[23] The lower box was mounted on a bearing frame, which ensures a lateral shearing to the load. The shear rate for all the testing was 1.25mm/min. The normal loads adopted for the study were 50 kPa, 75kPa, 100kPa, and 125kPa. The lower portion of the box was then filled with the prepared wet sand for a required weight and compacted in three layers with 110 blows. The compacted soil was weighed to ensure 95% relative compaction. IPB occupied the upper portion of the box with the joint width filled by jointing sand. The horizontal displacement, vertical displacement and load were recorded by two LVDTs along with a load cell attached to a data logger. Based on the measured load and displacement, peak shear strength is determined.

6.5 SUMMARY

The components of ICBP include the wearing layer such as IPB, bedding sand, and jointing sand along with the supporting layers base, subbase, and subgrade. The construction process of the supporting layers, gradation, and material properties are similar to that of the flexible pavement. The IPB and bedding sand are properly laid and compacted. The edge restraint needs to be properly fixed to resist the horizontal movement of ICBP and to enhance interlocking of blocks. The jointing sand needs to be filled between the blocks and well compacted using earth vibrator and roller. The performance of the laid ICBP can be better evaluated through various experiments such as plate load test, large scale direct shear test, LWD test, and BBD test. These tests aid in evaluating and monitoring the performance of the ICBP.

REFERENCES

1. IRC SP 63 (2018) Guidelines for the Use of Interlocking Concrete Block Pavement. *Indian Road Congress Special Publication*, New Delhi.
2. MoRTH (2013) Specifications for Roads and Bridge Works. *Ministry of Road Transport and Highways,* New Delhi.
3. IRC 36 (2010) Recommended Practice for Construction of Earth Embankments and Subgrade for Road Works. *Indian Road Congress Special Publication*, New Delhi.
4. IRC 37 (2018) Guidelines for the Design of Flexible Pavements. *Indian Road Congress*, New Delhi.

5. IRC SP 49 (2014) Guidelines for the Use of Dry Lean Concrete as Sub-Base for Rigid Pavement (as per IRC SP 49), Indian Road Congress, New Delhi.
6. Shackel, B. (1986) A Review of Research into Concrete Segmental Pavers in Australia Workshop on Interlocking Concrete Pavements. *Workshop on Interlocking Concrete Pavements*, Melbourne, Australia.
7. Mahapatra, G. and K. Kalita (2018) Effects of Interlocking and Supporting Conditions on Concrete Block Pavements. *Journal of The Institution of Engineers (India): Series A*, 99(1), 29–36.
8. Sarkar, H., C. Halder and T.L. Ryntathiang (2014) Behavior of Interlocking Concrete Block Pavement over Stone Dust Grouted Subbase. *International Journal of Advanced Structures and Geotechnical Engineering*, 3(1).
9. Raymond, S.R. (1984) Corps of Engineers Design Method for Concrete Block Pavements. *Second International Conference on Concrete Block Pavement,* Delft, The Netherlands.
10. Muraleedharan, T. and P.K. Nanda (1996) Laboratory and Field Study on Interlocking Concrete Block Pavement for Special Purpose Paving in India. *Fifth International Conference on Concrete Block Paving*, Tel Aviv, Israel.
11. Konard, R., P. WeUner and T. Gleitz (1994) The Behaviour of Dynamically Loaded Pavings. *Second International Conference on Concrete Block Pavement*, Osto, Norway.
12. Panda, B.C. and A.K. Ghosh (2002a) Structural Behavior of Concrete Block Paving. I: Concrete Blocks. *Journal of Transportation Engineering*, 128(2), 123–129. doi:10.1061/(ASCE)0733-947X(2002)128:2(123)
13. Lin, W., Y.H. Cho and I.T. Kim (2016) Development of Deflection Prediction Model for Concrete Block Pavement Considering the Block Shapes and Construction Patterns. *Advances in Materials Science and Engineering*, 2016. https://doi.org/10.1155/2016/5126436
14. Sharp, K.G., P.J. Armstrong and P.O. Morris (1982) The Performance of Two Interlocking Concrete Block Pavements. *Proceedings of Australian Road Research Board*, 11(2).
15. Ryntathiang, T.L., M. Mazumdar and B.B. Pandey (2006) Concrete Block Pavement for Low Volume Roads. *Eighth International Conference on Concrete Block Paving*, San Francisco, CA, USA.
16. Woodman, G.R. and A.R. Halliday (1992) The Performance of Concrete Block Surfacing on a Cement-Bound Base in Airfield Pavements. *Fourth International Conference on Concrete Block Pavement*, Auckland, New Zealand.
17. ASTM E2583 (2015b). Standard Test Method for Measuring Deflections with a Light Weight Deflectometer (LWD), *American Society for Testing and Materials,* Pennsylvania, United States.

18. Baba, T. and H. Yaginuma (2000) Evaluation of Bearing Capacity of Interlocking Block Pavement using HFWD. *Sixth International Conference on Concrete Block Paving*, Tokyo, Japan, pp. 61–70.
19. Adhikari, S., R. Burak, and S. Tighe (2009) Evaluation of Interlocking Concrete Pavement Crosswalks Through and Innovative Field Experiment, *Proceeding of, 9th International Conference on Concrete Block Paving*.
20. Park, H.M., S.H. Lee, S.A. Kwon, and Y.T. Kim (2018) Structural Evaluation of Block Pavements using Light Weight Deflectometer, *Twelfth International Conference on Concrete Block Pavement*, Seoul, Korea.
21. IRC 81 (1997) Guidelines for Strengthening of Flexible Road Pavements using Benkelman Beam Deflection Technique, *Indian Road Congress*. New Delhi.
22. ASTM D5321/D5321M (2019) Standard Test Method for Determining the Shear Strength of Soil-Geosynthetic and Geosynthetic-Geosynthetic Interfaces by Direct Shear. *American Society for Testing and Materials*, Pennsylvania, United States.
23. Arjun Siva Rathan, R.T., V. Sunitha., P. Murshidha., L. Janani., and V. Anusudha (2021) Experimental and Numerical Evaluation of the Parameters Influencing the Shear-Stress Behavior of Interlocking Paver Blocks – Bedding Sand Interface Using Large Scale Direct Shear Test. *Journal of Materials in Civil Engineerig*, 33(6), 04021104. https://ascelibrary.org/doi/full/10.1061/(ASCE)MT.1943-5533.0003724

Guidelines and Case Studies ICBP

7

7.1 INTRODUCTION

The design of ICBP follows the conventional flexible pavement design. The wearing layer of ICBP is designed based on different methodologies adopted in different countries. The design of ICBP mainly follows the equivalence factor of ICBP wearing layer to the flexible pavement wearing layer. Most of the countries follow their own methodology for the design of ICBP. However, most of the countries developed the design procedure of ICBP by modifying the existing design standards for flexible pavement. The base layer and subbase layer thickness was determined from the design charts used for the flexible pavement design. The application of the ICBP is not limited to highway pavements, but also extends its application to ports and airport pavements. The application of ICBP is increasing in mere days due to its better structural performance. The importance of the ICBP is better evaluated by measuring the performance of the ICBP in ports, highways and airports that have been laid before. The case studies revealed the fact that the ICBP is performing better in functional and structural characteristics.

DOI: 10.1201/9781003432371-7

7.2 GUIDELINE ADOPTED IN DIFFERENT COUNTRIES

7.2.1 Australia

The Concrete Masonry Association of Australia (CMAA) is a technical body for concrete pavers which provides specification and standards for the ICBP in Australia.[1] There are major design requirements for the determination of the thickness of ICBP base layers. The design thickness of unbound materials is based on the rutting criteria. The shape, laying pattern, and block thickness of the IPB is determined based on the catalogue provided in the guidelines which is shown in Table 7.1. The base course material should be provided to a compacted thickness range from 100mm to 200mm. The base thickness can be reduced by 20mm when the IPB of 80mm thickness is laid in herringbone bond for traffic up to 10^4 commercial vehicles. Well graded sand is preferred for the bedding sand. The moisture content to be maintained during compaction of bedding sand ranges from 4% to 8%. The joint width is maintained as 2mm to 5mm. The joint sand is preferred with sand passing 75 micron sieve. The guidelines aid in attaining the thickness of IPB shape, thickness, and laying pattern with respect to the traffic. The base layer is calculated based on the subgrade California Bearing ratio (CBR) and estimated commercial vehicles.

7.2.2 New Zealand

The Cement and Concrete Association of New Zealand and New Zealand Concrete Masonry Association formulated the design procedure for ICBP.[2]

TABLE 7.1 IPB shape thickness and laying pattern[1]

COMMERCIAL VEHICLES	*IPB SHAPE*	*THICKNESS (MM)*	*IPB LAYING PATTERN*
Up to 10^3	A, B or C	60	H,B or S
10^3 to 10^4	A	60	H only
	A, B or C	80	H,B or S
Over 10^4	A only	80	H only

Notes:
Herringbone = H, Basketweave = B, Stretcher = S
Type A = Interlocking on all sides, Type B = Interlocking on two sides
Type C = No interlocking on all sides.

TABLE 7.2 IPB geometric properties[2]

EQUIVALENT DESIGN AXLE	*IPB SHAPE*	*IPB THICKNESS (MM)*	*IPB LAYING PATTERN*
Up to 3 × 10^4	Any	60 or 80	H, B or S
3 × 10^4 or more	Capable of Herringbone	80	H only

Notes: Herringbone = H, Basket Weave = B, Stretcher = S.

The design is recommended for design traffic speed ranges from 50 kmph to 60 kmph. The design of the ICBP is based on Equivalent Design Axle (EDA) and subgrade strength in terms of CBR. The thickness of the IPB wearing layer is determined based on the catalogue provided in Table 7.2. The thickness, shape, and laying pattern of IPB are derived based on the design traffic considering a commercial vehicle with a laden weight greater than 3.5 Ton. Herringbone pattern is recommended for heavy traffic areas. The unbound base layer is determined from the design chart based on the subgrade CBR. The minimum granular base layer should be 50mm. The design chart of the base layer is available for IPB thickness of 60mm and 80mm. Based on the IPB thickness, the design chart will be chosen for the determination of unbound base layers. The guidelines aid in determining the wearing layer thickness and base layer thickness using two different design charts.

7.2.3 India

The Indian Road Congress (IRC) published the guidelines for the design of ICBP.[3] The design of ICBP follows the design catalogue developed based on the different traffic conditions and subgrade CBR. The design sections are provided directly for different traffic conditions. The ICBP section for light traffic pavements and industrial pavements is provided in Table 7.3. The block thickness used for light traffic is 60mm and for industrial roads is 100–120mm. The bedding compacted thickness ranges from 25 to 35mm. The bedding sand is compacted with the Optimum Moisture Content (OMC) of 6%. The maximum length of the IPB is restricted to 280mm. The water cement ratio for the manufacturing of IPB is maintained as 0.34 to 0.38. The minimum compressive strength of 35 MPa is suggested for IPB. In India, the design of ICBP section follows the design catalogue for different traffic conditions. The ICBP sections for different traffic in terms of million standard axle (MSA) is provided in Chapter 5, Table 5.1.

TABLE 7.3 ICBP section for light traffic and industrial pavements[3]

SL NO	*TRAFFIC CONDITION*	*SECTION DETAILS*
1.	Light traffic Pavements – Includes sidewalks, footpaths, cycle tracks and car parks	IPB – 60mm Bedding sand – 30±5mm Base course – 200mm
2.	Industrial Pavements – Container Yards, Port Wharf and warehouse roads	IPB – 100–120mm Bedding sand – 30±5mm Cement Base course – 300mm Granular Subbase – 300mm (150mm drainage layer)

TABLE 7.4 Tolerable deformations[4]

SL NO.	*TYPE OF ROADS*	*TOLERABLE DEFORMATION (MM)*
1.	Residential Street	10–15
2.	Rural Road	10
3.	Collector Street	7–12
4.	City Street	5–10
5.	Bus Stop	5

7.2.4 South Africa

The Concrete Manufacturers Association (CMA) is the technical body for the precast concrete products in South Africa.[4] CMA prepares technical reports and standards for ICBP. The design of the ICBP results in the determination of the thickness of IPB and thickness of the base layers. The increase in block thickness improves the performance of ICBP. The IPB thickness used for the pavement ranges from 50mm to 80mm. The block thickness is decided based on the traffic load. The block thickness of 50mm and 60mm is used for the ICBP for low traffic road pavements. The block thickness of 80mm is used for industrial application and 100mm thickness block is used for heavy loaded areas. The IPB should possess a minimum compressive strength of 25 MPa. The bedding sand layer should be provided to a compacted thickness of 25mm ± 10mm. The tolerable deflection for the heavy traffic loads is kept as 5mm. The thickness of the base layers are determined based on the design chart. The subgrade CBR and tolerable deformations are the required parameter to be considered for the determination of the base thickness. The tolerable deformation of different types of road categories are listed in Table 7.4. Three different design charts are provided in the guidelines based on 60mm, 80mm and 100mm block thickness. The thickness of the base layer is determined by considering the thickness of block, tolerable deformation based on different road categories and subgrade CBR.

7.2.5 United Kingdom

Interpave is the product association of British Precast Concrete Federation (BPCF) in the UK which promotes and provides technical specifications of ICBP.[5] The British standard for the design of ICBP follows BS 7533-101.[6] The Interpave has followed the guidelines of British Standard for the design of ICBP. The thickness of the IPB is determined based on the type of ICBP application ranging from 50mm to 100mm thickness. There are design procedures for the light traffic and heavy traffic loads. The thickness of the ICBP section for different category of roads is enlisted in Table 7.5. The categories are made based on the application of ICBP. The thickness of the IPB is determined based on the category of road. The laying course thickness for all types of roads are found to be 30mm. The paving thickness for the light traffic is 50mm and 60mm based on the type of application. For heavy duty pavements, the design thickness of ICBP is determined based on the traffic condition and the subgrade CBR. The thickness of the subbase and improvement layer is determined for CBR values ranging from 2% to 30%. The design life of ICBP is taken as 20 years. The guidelines aid in attaining the thickness of ICBP section based on light traffic and heavy traffic pavements for different subgrade strength.

7.2.6 United States

The Interlocking Concrete Paver Institute is a pioneering association intended for the promotion of the ICBP application worldwide and in US.[7] They had published technical specifications covering all aspects of ICBP. The ASCE 58-16 standard for the ICBP is developed in association with ICPI.[8] The design procedure developed by ICPI follows the AASHTO specification for flexible pavement design.[9] The soils are classified into eight different categories of soil based on the drainage and CBR values. The eight different soil categories are subdivided for drainage characteristics. The drainage behaviour is categorized into good, medium, and poor and the soils which fall under the different drainage conditions is provided in Table 7.6. The initial step in the design process of ICBP is the calculation of Traffic Index (TI) from the Equivalent Single Axle Load (ESAL). The ICBP section is directly determined from the design chart developed in the guidelines. The sample design chart of first three categories of road for different based layers are provided in Chapter 5, Table 5.4.

TABLE 7.5 ICBP section[5]

CATE GORY	*TYPICAL APPLICATIONS*	*SUBBASE THICKNESS MM CBR*					*ROAD BASE (MM)*	*LAYING COURSE (MM)*	*PAVING THICKNESS MINIMUM (MM)*
		<2%	3%	4%	5%	>6%			
I	Heavy Traffic – BS 7533-101	400	350	250	150	150	125	30	60
II	Adopted Highways and other roads, Car Parks with occasionally heavy traffic, Footpaths, overridden by vehicular traffic	350	300	225	150	150	0	30	50
IIIa	Pedestrian areas with occasional traffic	250	150	100	100	0	70	30	50
IIIb	Car parks and footways	300	250	175	100	100	0	30	50
IV	Private drives, paths, patio	200	150	125	100	0	0	30	50

TABLE 7.6 Drainage conditions for different soil categories[7,8]

QUALITY OF DRAINAGE	*TIME TO DRAIN*	*SOIL CATEGORIES*
Good	1 day	Boulders/cobbles, GW, SW, GP, SP
Fair	7 days	GM, SM, GC, SC
Poor	1 Month	GC, SC, ML, MI, CL, MH, CL, CH

The thickness of the IPB for all the traffic condition is kept at 80mm and the bedding sand thickness at 25mm. The minimum base thickness for granular layer is 200mm and the unconfined compressive strength of cement bound base is 4.5 MPa. The ICBP section can be selected from the input data such as subgrade strength, drainage condition and the traffic index calculated based on ESAL.

7.3 CASE STUDIES

7.3.1 Case Study I

Performance Reviews of Hong Kong International Airport and Yantian International Container[10]

The case study illustrates the performance of concrete block pavement laid in Hong Kong International Airport (HKIA) and Yantian International Container Terminals (YICT), in the People's Republic of China.

7.3.1.1 Hong Kong International Airport (HKIA)[10]

HKIA was built for six years and opened on July 1998. HKIA, during 1998 is one of the five busiest airports in the world handling 40.74 million passengers and 3.4 million tons of cargo. The HKIA aprons and service areas are constructed with ICBP which covers an area of approximately 500,000m^2. The ICBP is designed to withstand the aircraft weight up to 770 tons with a wheel load of 33 tons. The design of ICBP follows the conventional flexible airfield pavement design. The bituminous layer is replaced by the Interlocking Paver Block (IPB) and bedding sand. The wearing layer ICBP section consists of 80mm thick IPB and 20mm bedding sand. The geotextiles is placed above 175mm cement bound base to avoid the intrusion of bedding sand to the base layers. The joints are sealed to avoid the extrusion of jointing sand due to the jet engine blasts.

The visual inspection of the ICBP on the aircraft apron shows better performance and there is no evidence of pavement distress. The presence of fuel spillage is found on the surface of ICBP and found to be unaffected. The surface of ICBP is found to be flat without undulations however, a few isolated broken blocks are found in apron areas due to the impact load of aircraft. There is no indication of joint sand erosion from the joints between blocks. Minor deformation is found in the areas near to the terminal buildings and aprons. There are fault defects to a depth of 10 to 20mm but in an area of less than one square metre. There is no evidence of longitudinal rutting on the ICBP. The ICBP distress on airport pavement is witnessed to a great degree on areas around service trenches, pits, and manholes. The airport pavement constructed with ICBP proves to be a better performer and be able to withstand the heavy aircraft load with minimum maintenance.

7.3.1.2 Yantian International Container Terminals (YICT)[10]

YICT is one of the notable container ports in Shenzhen, Republic of China. A total of 1.3 million square metres of IPB is laid for handling 7.6 million TEU from 1998 to 2005. The design of ICBP is carried out for stacked containers constituting 24 tons point load. The shifting of the containers is carried out by eight-wheeled Rubber Tied Gantry Cranes (RTGC) weighing up to 220 tons with wheel loads of 27 tons/wheel. The other loads which involves the ICBP design for ports are forklifts and trucks. The impact of container trucks when compared to RTGC is found to be much lower. The ICBP in ports are laid with the IPB of 100mm thickness. The compacted thickness of the bedding sand beneath the IPB is kept to 25mm. The base course material used for ICBP is lean concrete mix to a thickness of 500mm and compressive strength of 10–20 MPa. Due to the availability of huge manpower, the blocks are laid by hand.

The visual inspection on the ports after eight years of construction confirms that ICBP performs well in both structural and functional characteristics. There is no evidence of adverse pavement failure on the surface which reduces the structural performance of ICBP. However, there are a few areas where the IPB are broken yet have not been isolated due to the impact load from the container trucks and RTGC. Being handled by a variety of goods, the debris are filled in between the block joints. The performance of ICBP is moderate at the locations such as test pits, drainage channels etc. due to minor settlement issues. In a few areas of access pathways, the rut depth is 300mm wide and 20mm. There are no major depressions on port pavements. Compared to the HKIA pavement YICT pavement performance is slightly lower. This is due to the fact that the entire ICBP section is constructed using machines in HKIA pavement and the manual laying is carried out in YICT. However, both the pavements perform well even after eight years of laying.

7.3.2 Case Study II

Three-year performance of concrete block pavements, under heavy abrasive caterpillar loading[11]

This case study describes the performance of interlocking concrete block pavements (CBP) exposed to heavy and abrasive caterpillar loading during their first three years of service. The construction of the test road section and apron area is completed in 1983 and opened for a heavy duty caterpillar vehicle in 1984. Based on the behaviour of CBP and their performance function with time, the condition of the pavement was analysed with respect to three types of traffic categories: (a) roads and aprons for pneumatic vehicles and for vehicles with rubber-padded caterpillar chains; (b) aprons for parking and maintenance of heavy vehicles with exposed metallic chains; and (c) access roads for fast heavy vehicles with exposed metallic chains. The concrete block pavement structure proposed for design was interlocking concrete block layer (UNI type) of thickness 10 cm. Sand bedding layer of thickness 4 cm. Subbase course (type A) of thickness 26 cm. The physical requirements for the concrete blocks included the compressive strength, bending strength and abrasion resistance test according to Israel Standards Nos. IS-26, IS-106 and IS-108. The UNI-type HD10 blocks were produced in a newly erected plant with a line specially designed and constructed for concrete block production.

From the performance study over three years, it is witnessed that the ICBP is a durable pavement built to withstand the slow and fast heavy caterpillar with rubber padding chains. However, the metallic chained heavy caterpillar vehicles prove abrasion and upper corner breakage of pavements, which mainly depends on traffic intensity at each location. The abrasion and corners breakage gradually developed with time, producing a widening of the upper joints between blocks. The weight of the test vehicle is around 50 tons. After three years of service, the thickness of the block still maintains its interlocking effect; thus, the pavement remains stable and monolithic, without any settlement, rutting or horizontal movement of the block. The fast movement of heavy caterpillar vehicles with exposed metallic chains in access roads caused excessive abrasion and breakage of entire blocks at the end section of access roads after three years of service. This abrasion and breakage created the loss of the interlocking potential of the pavement, causing the loss of bearing capacity and developing settlement, rutting and local failure of the entire block pavement. The local total failure of the pavement could have been prevented by the replacement in time of the small abrasive spot with new blocks. This act of maintenance is quite easy and cheap when performed in time and represents one of the major advantages of concrete block paving. Based on the three-year performance, CBP was found to be unsuitable for operating heavy caterpillar vehicles with exposed metallic chains. On the other hand, CBP is quite suitable and durable for any heavy pneumatic vehicles and heavy caterpillar vehicles with rubber-padded chains. With proper pavement design, block production and construction procedure, an ICBP can provide a durable pavement to support both the slow and fast traffic of heavy caterpillar vehicles.

7.4 SUMMARY

The ICBP design procedure is different for different countries. Most of the countries followed the design procedure of flexible pavement and the wearing layer is altered using equivalency factor. It is also witnessed that most of the countries followed empirical design approach for attaining the ICBP section. The performance of ICBP is calibrated from the field testing of previously laid ICBP. The case studies on the application of ICBP in road, airport, and port proves that ICBP is a sustainable pavement which can withstand high traffic load. The ICBP is found to be a high performance pavement which can be used for heavy traffic areas; however, the application is restricted to the areas where the design speed of the vehicle is greater than 50 or 60 kmph.

REFERENCES

1. CMAA (2014) Concrete Segmental Pavements – Detailing Guide. *Concrete Manufacturing Association of Australia*, Australia.
2. CMA (1988) Interlocking Concrete Block Road Pavements – IB 67. *Concrete Masonry Association*, New Zealand.
3. IRC SP 63 (2018) Guidelines for the Use of Interlocking Concrete Block Pavement. *Indian Road Congress Special Publication*, New Delhi.
4. CMA (2004) Concrete Block Paving Book 2: Design Aspects. *Concrete Manufacturers Association*, South Africa.
5. Interpave (2010) The Precast Concrete Paving & Kerb Association. *British Precast Concrete Federation Ltd*, United Kingdom.
6. BS 7533-101 (2021) Pavements Constructed with Clay, Concrete or Natural Stone Paving Units: Part 101. Code of Practice for the

Structural Design of Pavements Using Modular Paving Units. *British Standards Institution*, United Kingdom.

7. ICPI (2020) Structural Design of Interlocking Concrete Pavement for Roads and Parking Lots. *Interlocking Concrete Pavement Institute*, Virginia, United States.
8. ASCE 58-16 (2016) Structural Design of Interlocking Concrete Pavement for Municipal Streets and Roadways. *American Society of Civil Engineers*, Virginia, United States.
9. AASHTO (1993) Guide for Design of Pavement Structures. *American Association of State Highway and Transportation Officials,* United States.
10. Claudia Yun Kang (2006) Performance Reviews of Hong Kong International Airport and Yantian International Container. *8th International Conference on Concrete Block Paving*, San Francisco, California, USA.
11. Ishai, I., and M. Livneh (1988) Three-Year Performance of Concrete Block Pavements Under Heavy Abrasive Caterpillar Loading. *3rd International conference on Concrete Block Paving*, 361–370, Rome, Italy.

Index

Note: Figures are indicated by *italics*. Tables are indicated by **bold**.

F

G

H

I

J

L

M

N

O

P